T0320141

Breakthroughs in Decision Science and Risk Analysis

Wiley Essentials in
Operations Research and Management Science

Founding Series Editor
James J. Cochran, The University of Alabama

Wiley Essentials in Operations Research and Management Seience is a unique collection of international research written by recognized experts and features both state-of-the-art developments in the field and classical methods that are appropriate for researchers, practitioners, policy makers, academicians, and students alike. Inspired by the *Wiley Encyclopedia of Operations Research and Management Science* (EORMS), this authoritative series covers topics of current research and practical interest and has a global emphasis on specific and key areas of study within the diverse field of operations research and management science. This broad collection of books provides wide-ranging and complete coverage within individual books and unifies the most important and key advances in the theory, methods, and applications within a specific area of study. Each book in the series successfully upholds the goals of EORMS by combining established literature and newly developed contributions written by leading academics, researchers, and practitioners in a comprehensive and accessible format. The result is a succinct reference that unveils modern, cutting-edge approaches to acquiring, analyzing, and presenting information across various subject areas.

Breakthroughs in Decision Science and Risk Analysis

Edited by

Louis Anthony Cox, Jr.

Cox Associates
NextHealth Technologies
University of Colorado-Denver
Denver, CO, USA

Library of Congress Cataloging-in-Publication Data:

Breakthroughs in decision science and risk analysis / [edited by] Louis Anthony Cox, Jr.
     pages   cm
   Includes bibliographical references and index.
   ISBN 978-1-118-21716-0 (cloth)
1. Decision making.   2. Risk assessment.   I. Cox, Louis A.
   T57.95.B74 2014
   658.4′03–dc23

                                                          2014012640

Printed in the United States of America.

10   9   8   7   6   5   4   3   2   1

Contents

Foreword

The *Wiley Encyclopedia of Operations Research and Management Science* (*EORMS*), published in 2011, is the first multi-volume encyclopedia devoted to advancing state-of-the-art applications and principles of operations research and management science. *EORMS* is available online and in print and serves students, academics, and professionals as a comprehensive resource on the field.

The articles published in *EORMS* provide robust summaries of the many topics and concepts that comprise operations research and management science. However, many readers need additional access to greater details and more thorough discussions on a variety of topics. In turn, we have created the *Wiley Essentials in Operations Research and Management Science* book series. Books published in this series allow invited expert authors and editors to expand and extend the treatment of topics beyond *EORMS* and offer new contributions and syntheses.

I am delighted to introduce *Breakthroughs in Decision Science and Risk Analysis* as the inaugural book in this series. It exemplifies how individual books will meet the goals of the series, setting a high standard for later volumes. Dr. Louis Anthony (Tony) Cox, Jr., the editor of *Breakthroughs in Decision Science and Risk Analysis*, has assembled a collection of renowned authors who contribute exciting syntheses and advances to an area that is critically important in both applications and research. This book is unique and important because it focuses on recent advances and innovations in decision analysis with an emphasis on topics that are not traditionally found in the decision analysis literature. I am confident that this book will be extremely useful to the operations research and management science community as well as to readers from economics, engineering, business, psychology, finance, environmental sciences, public policy, forestry, political science, health and medicine, education, and other social and applied sciences. We hope that you will enjoy it, and we welcome comments and suggestions for our new *Wiley Essentials in Operations Research and Management Science* series.

James J. Cochran
Professor of Applied Statistics and the
Rogers-Spivey Faculty Fellow
Founding Series Editor of Wiley Essentials in
Operations Research and Management Science
University of Alabama

Preface

Decision and risk analysis can be exciting, clarifying, and fairly easy to apply to improve high-stakes policy and management decisions. The field has been undergoing a renaissance in the past decade, with remarkable breakthroughs in the psychology and brain science of risky decisions; mathematical foundations and techniques; integration with learning and pattern recognition methods from computational intelligence; and applications in new areas of financial, health, safety, environmental, business, engineering, and security risk management. These breakthroughs provide dramatic improvements in the potential value and realism of decision science. However, the field has also become increasingly technical and specialized, so that even the most useful advances are not widely known or applied by the general audience of risk managers and decision-makers who could benefit most from them. This book explains key recent breakthroughs in the theory, methods, and applications of decision and risk analysis. Its goal is to explain them in enough detail, but also with a simple and clear enough exposition, so that risk managers and decision-makers can understand and apply them.

There are several target audiences for this book. One is the operations research and management science (OR/MS) community. This overlaps with the audience for the *Wiley Encyclopedia of Operations Research and Management Science* (EORMS), including OR/MS professionals–academic and industry researchers, government organizations and contractors, and decision analysis (DA) consultants and practitioners. This book is also designed to appeal to managers, analysts, and decision and policy makers in the applications areas of financial, health and safety, environmental, business, engineering, and security risk management. By focusing on breakthroughs in decision and risk science that can significantly change and improve how we make (and learn from) important practical decisions, this book aims to inform a wide audience in these applied areas, as well as provide a fun and stimulating resource for students, researchers, and academics in DA and closely linked academic fields of psychology, economics, statistical decision theory, machine learning and computational intelligence, and OR/MS.

In contrast to other recent books on DA, this one spends relatively little space on "classical" topics such as the history of DA; the structure of decision problems in terms of acts, states, and consequences; and extensions of the traditional DA paradigm, such as fuzzy DA or multi-criteria decision-making. Instead, the book is devoted to explaining and illustrating genuine *breakthroughs* in decision science—that is, developments that depart from, or break with, the standard DA paradigm in fundamental ways and yet have proved promising for leading to even more valuable insights and decision recommendations in practical applications. These breakthroughs include methods for deciding what to do when decision problems are incompletely known or described, useful probabilities cannot be specified, preferences and value trade-offs are uncertain, future preferences may conflict with present ones, and "model uncertainty" about the cause-and-effect relation between choices and their probable consequences runs too deep to permit any single decision support model, or small set of models, to be highly credible. In addition to explaining the most important new ideas for coping with such realistic challenges, the chapters in this book will show how these techniques are being used to dramatically improve risk management decisions in a variety of important applications, from finance to medicine to terrorism. This emphasis on *new ideas that demonstrably work better than older ones*, rather than primarily on expositions of and advances in traditional decision and risk analysis, is the essential unique contribution of this work. In addition, a pedagogical emphasis on *simple, clear exposition* (accessible to readers at different technical levels, with a minimum of mathematical notation and technical jargon), and *important practical applications* should help to broaden the practical value of the chapters that follow in making important advances in decision and risk analysis useful to readers who want to learn about them and apply them to important real-world decisions.

Louis Anthony Cox, Jr.
January 2015

Contributors

Ali E. Abbas
Industrial and Systems Engineering,
Center for Risk and
Economic Analysis of Terrorism Events
Sol Price School of Public Policy
University of Southern California
Los Angeles, CA, USA

Marco Better
OptTek Systems, Inc.
Boulder, CO, USA

Vicki M. Bier
Department of Industrial and Systems Engineering
University of Wisconsin-Madison
Madison, WI, USA

Louis Anthony (Tony) Cox, Jr.
Cox Associates
NextHealth Technologies
University of Colorado-Denver
Denver, CO, USA

Fred Glover
OptTek Systems, Inc.
Boulder, CO, USA

Andrea C. Hupman
Department of Industrial and Enterprise Systems Engineering
College of Engineering
University of Illinois at Urbana-Champaign
Urbana-Champaign, IL, USA

Gary Kochenberger
Business Analytics
School of Business
University of Colorado-Denver
Denver, CO, USA

Philip Leclerc
Department of Statistical Sciences and Operations Research
Virginia Commonwealth University
Richmond, VA, USA

Jason Merrick
Department of Statistical Sciences and Operations Research
Virginia Commonwealth University
Richmond, VA, USA

Robert Olsen
Deloitte Consulting
Center for Risk Modeling and Simulation
Washington, DC, USA

Edward C. Rosenthal
Department of Marketing and Supply Chain Management
Fox School of Business
Temple University
Philadelphia, PA, USA

Ilya O. Ryzhov
Robert H. Smith School of Business
University of Maryland
College Park, MD, USA

Sinan Tas
Information Sciences and Technology
Penn State-Berks
Reading, PA, USA

Hristo Trenkov
Deloitte Consulting
Center for Risk Modeling and Simulation
Washington, DC, USA

Chen Wang
Department of Industrial Engineering
Tsinghua University
Beijing, P.R., China

CHAPTER 1

Introduction: Five Breakthroughs in Decision and Risk Analysis

Louis Anthony (Tony) Cox, Jr.
*Cox Associates, NextHealth Technologies,
University of Colorado-Denver, Denver, CO, USA*

This book is about breakthroughs in decision and risk analysis—new ideas, methods, and computational techniques that enable people and groups to choose more successfully when the consequences of different choices matter, yet are uncertain. The twentieth century produced several such breakthroughs. Development of subjective expected utility (SEU) theory combined with Bayesian statistical inference as a model of ideal, rational decision-making was among the most prominent of these. Chapter 2 introduces SEU theory as a point of departure for the rest of the book. It also discusses more recent developments—including prospect theory and behavioral decision theory—that seek to bridge the gap between the demanding requirements of SEU theory and the capabilities of real people to improve their decision-making. Chapters 5 and 8 address practical techniques for improving risky decisions when there are multiple objectives and when SEU cannot easily be applied, either because of

Breakthroughs in Decision Science and Risk Analysis, First Edition.
Edited by Louis Anthony Cox, Jr.

uncertainty about relevant values, causal models, probabilities, and consequences; or because of the large number and complexity of available choices.

HISTORICAL DEVELOPMENT OF DECISION ANALYSIS AND RISK ANALYSIS

Perhaps the most audacious breakthrough in twentieth-century decision analysis was the very idea that a single normative theory of decision-making could be applied to all of the varied risky decisions encountered in life. It may not be obvious what the following problems, discussed in ensuing chapters, have in common:

- *Investment decisions:* How should investors allocate funds across investment opportunities in a financial portfolio? (Chapter 3)

- *Operations management decisions:* How should a hospital emergency room be configured to make the flow of patients as easy and efficient as possible? How should an insurance company staff its claims-handling operations? (Chapter 3)

- *Inventory management and retail decisions:* How much of an expensive, perishable product should a business buy if demand for the product is uncertain? (Chapter 4)

- *Trial evaluation and selection decisions:* How much trial, testing, and comparative evaluation should be done before selecting one of a small number of costly alternatives with uncertain consequences, for example, in choosing among alternative new public policies, consumer or financial products, health care insurance plans, research and development (R&D) projects, job applicants, supply contracts, locations in which to drill for oil, or alternative drugs or treatments in a clinical trial? (Chapter 4)

- *Adversarial risk management decisions:* How should we model the preferences and likely actions of others, in order to make effective decisions ourselves in situations where both their choices and ours affect the outcomes? (Chapters 2, 5, 9, and 10)

- *Regulatory decisions:* When experimentation is unethical or impractical, how can historical data be used to estimate and compare the probable consequences that would be caused by alternative choices, such as revising versus maintaining currently permitted levels of air pollutants? (Chapter 6)

- *Learning how to decide in uncertain environments:* Suppose that not enough is known about a system or process to simulate its behavior. How can one use well-designed trial-and-error learning to quickly develop high-performance decision rules for deciding what to do in response to observations? (Chapters 4 and 7)

- *Medical decision-making:* How should one trade off the ordinary pleasures of life, such as consumption of sugar-sweetened drinks, against the health risks that they might create (e.g., risk of adult-onset diabetes)? More generally, how can and should individuals make decisions that affect their probable future health states in ways that may be difficult to clearly imagine, evaluate, or compare? (Chapter 8)

That the same basic ideas and techniques might be useful for decision-making in such very different domains is a profound insight that might once have excited incredulity among experts in these fields. It is now part of the canon of management science, widely taught in business schools and in many economics, statistics, and engineering programs.

Decision analysis views the "success" of a decision process in terms of the successes of the particular decisions that it leads to, given the information (usually incomplete and possibly incorrect or inconsistent) that is available when decisions must be made. The "success" of a single choice, in turn, can be assessed by several criteria. Does it minimize expected post-decision regret? Is it logically consistent with (or implied by) one's preferences for and beliefs about probable consequences? In hindsight, would one want to make the same choice again in the same situation, if given the same information? The giants of twentieth-century decision theory, including Frank Ramsey in the 1920s, John von Neumann in the 1940s, and Jimmy Savage in the 1950s, proved that, for perfectly rational people (*homo economicus*) satisfying certain mathematical axioms of coherence and consistency (i.e., complete and transitive preference orderings for outcomes and for probability distributions over outcomes), all of these criteria prescribe the same choices. All imply that a decision-maker should choose among risky prospects (including alternative acts, policies, or decision rules with uncertain consequences) as if she were maximizing subjective expected utility (SEU). Chapters 2 and 7 introduce SEU theory and some more recent alternatives. Decision-making processes and environments that encourage high-quality decisions as judged by one of these criteria will also promote the rest.

However, real people are not perfectly rational. As discussed in Chapter 2, *homo economicus* is a fiction. The prescriptions of decision theory are not necessarily easy to follow. Knowing that SEU theory, the long-reigning gold standard for rational decision-making, logically implies that one *should* act as if one had coherent (e.g., transitive) preferences, and clear subjective probabilities are cold comfort to people who find that they have neither. These principles and limitations of decision theory were well understood by 1957, when Duncan Luce and Howard Raiffa's masterful survey *Games and Decisions* explained and appraised much of what had been learned by decision theorists, and by game theorists for situations with multiple interacting decision-makers. Chapter 2 introduces both decision theory and game theory and discusses how they have been modified recently in light of insights from decision psychology and behavioral economics.

During the half-century after publication of *Games and Decisions*, a host of technical innovations followed in both decision analysis and game theory. Decision tree analysis (discussed in Chapters 8 and 10) was extended to include Monte Carlo simulation of uncertainties (see Chapter 3). Influence diagrams were introduced that could represent large decision problems far more compactly than decision trees, and sophisticated computer science algorithms were created to store and solve them efficiently. Methods of causal analysis and modeling were developed to help use data to create risk models that accurately predict the probable consequences of alternative actions (see Chapter 6). Markov decision processes for dynamic and adaptive decision-making were formulated, and algorithms were developed to adaptively and robustly optimize decision rules under uncertainty (see Chapter 7). SEU theory was generalized, e.g., to allow for robust optimization with ambiguity aversion when probabilities are not well known. Practical constructive approaches were created for structuring and eliciting probabilities and utilities, as discussed and illustrated in Chapters 5, 8, and 10.

These technical developments supported a firmly founded discipline of applied decision analysis, decision aids, and decision support consulting. The relatively new discipline of applied decision analysis, developed largely from the 1960s on, emphasized structuring of decision problems (especially, identifying and solving the right problem(s)); clearly separating beliefs about facts from values and preferences for outcomes; eliciting or constructing well-calibrated probabilities and coherent utilities; presenting decision recommendations, together with sensitivity and uncertainty analyses, in understandable ways that decision-makers find useful; assessing value of information and optimal timing of actions; and deliberate, careful learning from results, for both individuals and organizations. The 1976 publication of the landmark *Decisions with Multiple Objectives: Preferences and Value Tradeoffs* by Ralph Keeney and Howard Raiffa summarized much of the state of the art at the time, with emphasis on recently developed multiattribute value and utility theory and methods. These were designed to allow clearer thinking about decisions with multiple important consequence dimensions, such as costs, safety, profitability, and sustainability. Chapters 5, 8, and 10 review and illustrate developments in elicitation methods and multiattribute methods up to the present.

While decision analysis was being developed as a prescriptive discipline based on normative theory (primarily SEU theory), an increasingly realistic appreciation of systematic "heuristics and biases" and of predictable anomalies in both laboratory and real-world decision-making was being developed by psychologists such as Amos Tversky, Daniel Kahneman, Paul Slovic, and Baruch Fischhoff, and by many other talented and ingenious researchers in what became the new field of behavioral economics. Chapter 2 introduces these developments. Striking differences between decision-making by idealized, rational thinkers (*homo economicus*) and by real people were solidly documented and successfully replicated by different teams of investigators. For example, whether cancer patients and their physicians preferred one risky treatment procedure to another might be changed by presenting risk information as the probability of survival for at least 5 years instead of as the probability of death within 5 years—two logically equivalent descriptions (gain faming vs. loss framing) with quite different emotional impacts and effects on decisions. Many of these developments were reflected in the 1982 collection *Judgment under Uncertainty: Heuristics and Biases*, edited by Kahneman, Slovic, and Tversky. Chapter 10 summarizes key insights from the heuristic-and-biases literature in the context of eliciting expert judgments about probabilities of adversarial actions.

Twenty-five years later, the 2007 collection *Advances in Decision Analysis: From Foundations to Applications*, edited by Ward Edwards,

Ralph Miles, and Detlof von Winterfeldt, took stock of the thriving and increasingly well-developed field of decision analysis, which now integrated both normative (prescriptive) theory and more descriptively realistic considerations, e.g., using Kahneman and Tversky's prospect theory. This collection looked back on decades of successful developments in decision analysis, including the field's history (as recalled by founding luminaries, including Ron Howard of Stanford and Howard Raiffa of the Harvard Business School), surveys of modern progress (including influence diagrams, Bayesian network models, and causal networks), and important practical applications, such as to engineering and health and safety risk analysis, military acquisitions, and nuclear supply chain and plutonium disposal decisions.

OVERCOMING CHALLENGES FOR APPLYING DECISION AND RISK ANALYSIS TO IMPORTANT, DIFFICULT, REAL-WORLD PROBLEMS

Despite over five decades of exciting intellectual and practical progress, and widespread acceptance and incorporation into business school curricula (and into some engineering, statistics, mathematics, and economics programs), decision analysis has limited impact on most important real-world decisions today. Possible reasons include the following:

• *Many real-world problems still resist easy and convincing decision-analytic formulations.* For example, a dynamic system with random events (i.e., patient arrivals, departures, and changes in condition in a hospital ward) with ongoing opportunities to intervene (e.g., by relocating or augmenting staff to meet the most pressing needs) cannot necessarily be represented by a manageably small decision tree, influence diagram, Markov decision process, or other tractable model—especially if the required transition rates or other model parameters are not known and data from which to estimate them are not already available. Chapters 3, 4, 6, and 7 present breakthroughs for extending decision and risk analysis principles to such realistically complex settings. These include increasingly well-developed *simulation–optimization* methods for relatively well-characterized systems (Chapter 3) and *adaptive learning*, statistical methods for estimating causal relations from data (Chapter 6), model ensemble, and robust optimization methods for settings where not enough is known to create a trustworthy simulation model (Chapters 4 and 7).

- *It has often not been clear that individuals and organizations using formal decision analytic models and methods outperform and outcompete those who do not.* Chapters 4, 6, and 7 emphasize methods for causal analysis, adaptive learning, and empirical evaluation and comparison of alternative choices. These methods can help decision-makers make choices that demonstrably outperform (with high statistical confidence) other available choices in a wide variety of practical applications.

- *While decision analysts excel at distinguishing clearly between matters of fact and matters of preference, real-world decision-makers often prefer to fall back on judgments that conflate the two,* perhaps feeling that no matter what academic theory may say, such holistic judgments give more satisfactory and trustworthy recommendations than calculations using hypothetical (and not necessarily clearly perceived or firmly believed) subjective utilities and probabilities. (This tendency may perhaps explain some of the popularity of simplistic decision aids that use ill-defined concepts, such as "relative importance" of goals or attributes, without clear definition of what "relative importance" means and of how it should reflect interdependencies.) Too often, there is simply no satisfactory way to develop or elicit credible, defensible, widely shared probabilities and utilities for situations or outcomes that are novel, hard to imagine, or controversial. Chapters 5 and 8–10 discuss innovations for alleviating this problem with new methods for eliciting and structuring utilities, value trade-offs, and probabilistic expert beliefs.

- *Most real-world decisions involve multiple stakeholders, influencers, and decision-makers,* but SEU is preeminently a theory for single-person decisions. (Extensions of SEU to "social utility" for groups, usually represented as a sum of individual utilities, can certainly be made, but an impressive list of impossibility theorems from collective choice theory establish that *homo economicus* will not necessarily provide the private information needed for collective choice mechanisms to produce desirable, e.g., Pareto-efficient, outcomes.) Less theoretically, the notorious Prisoner's Dilemma, discussed in Chapter 2, illustrates the tension between individual and group rationality principles. In the Prisoner's Dilemma, and in many other situations with externalities, individuals who make undominated or otherwise individually "rational" choices will thereby collectively achieve Pareto-dominated outcomes (a hallmark of collectively suboptimal choice), meaning that everyone would have been better off if all had made different (not "rational") choices. Chapter 2 discusses both classical game theory and its behavioral modifications to better apply to real people, who often cooperate far better than theories for merely "rational" individuals would predict. Chapters 9 and 10 consider applications of game theory and alternatives for defending electrical grids (Chapter 9) and other targets (Chapter 10) against terrorists or other adversaries, including natural disasters in Chapter 9.

In addition to these major conceptual challenges, there are also purely technical challenges for making decision-analytic principles more widely applicable. For example, decision trees (Chapter 5) are well suited to model (and if necessary simulate) alternative possible sequences of events and decisions when there are only a few of each. However, they are far from ideal when the number of choices is large or continuous.

Example: Searching for a Hidden Prize

Suppose that a prize is hidden in one of 100 boxes, and that the cost of opening each box to see whether the prize is in it, as well as the prior probability that it is, are known (say, $c(j)$ to open box j, which has prior probability $p(j)$ of containing the prize). Then in what order should the boxes be opened to minimize the expected cost of finding the prize? (This is a very simple model for sequential search and R&D problems.) It would clearly be impracticable to create a decision tree describing the 100! possible orders in which the boxes might be opened. Yet, it is easy to determine the optimal decision. Simple optimization reasoning establishes that the boxes should be opened in order of descending probability-to-cost ratio (since interchanging the order of any two boxes that violate this rule can readily be seen to reduce expected cost).

The remainder of this book explains and illustrates breakthrough methods to help real people make real decisions better. It presents ideas and methods that the authors and editors believe are mature enough to be highly valuable in practice and that deserve to be more widely known and applied. The starting point, developed in Chapter 2, is a candid acknowledgment that:

- SEU and classical Bayesian decision analysis together provide a logically compelling model for how individual decisions ideally should be made; but
- Real people have not evolved to be always capable of producing (or agreeing on) the crisply specified, neatly decoupled subjective probabilities and utilities that are required (and implied) by SEU.

Instead, decision-makers in real-world organizations and institutions typically have access to some imperfect data and knowledge bearing on the causal relations between alternative choices and their probable consequences. From this partial information, and through processes of deliberation and analysis, they may construct preferences for alternative actions, with supporting rationales that are more or less convincing to themselves and to others.

The following five breakthroughs, explained and illustrated in the chapters that follow, can help to understand and improve these decision processes.

Breakthrough 1: Behavioral Decision Theory and Game Theory

It is now known that different neural pathways and parts of the brain are activated by different aspects of risky prospects, such as probabilities versus amounts of potential perceived gains or losses; immediate versus delayed consequences; positive versus negative emotional affects of cues used in describing them; trust versus suspicion of others involved in joint decisions; and moral reactions to risks and to imposing risks on others. To a useful first approximation, heart and head (or, more formally, "System 1" and "System 2," referring to the quick, intuitive and slower, more cognitive aspects of decision-making, respectively) may disagree about what is best to do, using different parts of the brain to evaluate alternatives and to arrive at these conclusions. Chapter 2 further describes these aspects of the divided decision-making self, situating the problem of wise decision-making (as well as moral and social judgment-making) in the context of competing decision pathways and emphasizing the need to use both emotional and intuitive judgments and rational calculations in making effective decisions. Preferences, judgments, and beliefs are often transient and contingent on context and cues (some of which may be logically irrelevant) when they are elicited.

Assessing a single coherent utility function for a person whose preferences arise from multiple cues and competing pathways may not yield a reliable basis for prescribing what to do if the goal is to minimize post-decision regret.

Recognizing these realities of human nature and behavior motivates *behavioral decision theory* and *behavioral game theory*. These address how to use well-placed "nudges," design of occupational and consumer environments, and other practical methods to help real people make better decisions while taking into account their heuristics, biases, inconsistencies, limited attention span and will-power, irrational altruism, moral aspirations, perceptions of fairness, desire for social approval, and other realities of motivation and behavior. "Better" decisions can no longer necessarily be defined as those that are consistent with SEU axioms for preferences and beliefs, as assessed at a given moment and in a given context. What constitutes desirable decision-making must be defined afresh when preferences and beliefs are seen as being constructed on the fly and subject to motivated and self-serving reasoning, wishful thinking, salience of cues and their emotional affects, priming by context, and other biases. For example, "good" choices might be defined as those that are consistent with the guidance or principles (more formally, with the if-then "decision rules" mapping available information to available actions) that one would ideally want one's self to follow, if one were given time, resources, and ability to develop such principles outside the context of short-run distractions, passions, and temptations. Such reflective and reflexive considerations, which have a long tradition in deontological, utilitarian, and virtue ethics, are gaining new currency and an applied focus through behavioral decision theory and game theory. The core breakthrough in Chapter 2 is the insight that advice on how to make decisions, to be most useful, should be rooted to an understanding of real human behavior and realistic possibilities for changing it.

Breakthrough 2: Simulation–Optimization of Risk Models

For decades, one vision of applied decision analysis has been that knowledge and information about the system or situation that a decision-maker seeks to influence via her decisions should be represented in an explicit *risk model* relating decisions (controllable inputs) and uncertainties (e.g., modeled as random inputs from the environment, not controlled by the decision-maker) to resulting probabilities of different outputs (consequences). If expected utility, or any other "objective function" whose expected value is to be maximized, is used to evaluate the probabilities of consequences induced by alternative choices of controllable inputs, then the decision problem of selecting those inputs can be decomposed into the following two technical tasks:

1. *Simulate* output (consequence) probability distributions, for any choice of inputs; and

2. *Optimize* inputs, that is, identify a combination of input values to produce the most desirable probability distribution of outputs, as evaluated by the objective function (e.g., expected utility).

If the risk model and simulation–optimization process can be trusted to model adequately the real system or situation of interest and to automatically find the best combination of controllable inputs to choose for that system, then the decision-maker is freed to focus on specifying the controllable inputs and the objective function to be used in evaluating results. Appropriate subject matter experts and modelers can focus on developing and validating the risk model describing the probabilities of consequences caused by different choices of controllable inputs (together with the uncontrollable ones chosen by "nature" or by others). The simulation–optimization engine can handle the details of solving for the best choice of inputs, much as software products such as the Excel Solver or Wolfram Alpha can solve simper problems, freeing users to focus on model development and input specification.

Example: Optimal Level of R&D

Suppose that a pharmaceutical company can invest in investigating any of a large number of new leads (molecules and molecular signaling pathways) in parallel for developing a drug to treat a disease in patients with a certain genotype. Each lead costs $5M to investigate, and each has a probability 0.1 of proving successful within 5 years. The value of a success within 5 years is $100M. The company must decide how many leads to investigate in parallel. What number of leads should be investigated to maximize the expected profit? This objective function, in units of millions of dollars, is given by the formula:

$$\text{expected profit} = \text{probability of success} \times \$100\text{M}$$
$$- (\text{number of parallel investigations}) \times \$5\text{M}$$
$$= \left(1 - (1-p)^N\right) \times 100 - 5N,$$

where N is the number of leads investigated—the decision variable in this problem—and p is the probability of success for each investigated lead. (The probability that all N investigated leads are unsuccessful is $(1-p)^N$, and hence the probability of success for the whole effort, that is, the probability that not all fail, is $(1-(1-p)^N)$.) The expected profit-maximizing value of N can readily be found in this simple example by searching over a range of values of N. For example, for those familiar with R, the following code generates Fig. 1.1: N=c(1:20); EMV=100 ×(1−0.9^N)−5×N; plot (N, EMV). The number of leads that maximizes the objective function is $N=7$.

Now, suppose that the problem were more complicated, with unequal success probabilities and different costs for the different leads, and with annual budgets and other resource constraints limiting the number of projects (investigations) that could be undertaken simultaneously. Then instead of searching for the best solution over a range of values for N, it would be necessary to search a more complicated space of possible solutions, consisting of all subsets of projects (lead investigations) that can be investigated simultaneously (i.e., that satisfy the budget and resource constraints). If, in addition, the objective function could not easily be described via a formula, but instead had to be estimated by simulating many realizations of the uncertain quantities in the model for each choice of inputs, then efficient search and evaluation of different input combinations might become important, or even essential, for finding a good solution to the decision problem. Simulation-optimization provides technical methods for efficiently searching complex sets of feasible decisions, performing multiple simulation-based evaluations of alternative combinations of controllable inputs to identify those that (approximately) optimize the user-specified objective function.

The vision of decision-making as optimization of an appropriate objective function subject to constraints, corresponding to a model of how choices affect consequence probabilities, is fundamental in economics, operations research, and optimal control engineering (including stochastic, robust, and adaptive control variations). However, to make it practical, both the simulation and the optimization components must be

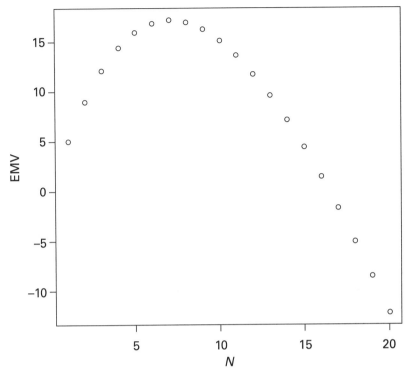

Figure 1.1   Plot of expected monetary value (EMV) for profit vs. N in the R&D example. Choosing $N = 7$ parallel projects (leads to investigate) maximizes expected net return.

well enough developed to apply to realistically complex systems. This is an area in which huge strides have been made in the past two decades, amounting to a breakthrough in decision problem-solving technology. Sophisticated stochastic simulation techniques (e.g., Gibbs sampling, more general Markov chain Monte Carlo (MCMC) techniques, Latin Hypercube sampling, importance sampling, and discrete-event simulation methods) and sophisticated optimization methods that work with them (e.g., evolutionary optimization, simulated annealing, particle filtering, tabu search, scatter search, and other optimization meta-heuristics) are now mature. They have been encapsulated in user-friendly software that presents simple interfaces and insightful reports to users, who do not need to understand the details of the underlying algorithms. For example, the commercially available OptQuest simulation-optimization engine discussed in Chapter 3 is now embedded in software products such as Oracle, Excel, and Crystal Ball. Its state-of-the-art optimization meta-heuristics make it practical to easily formulate and solve decision problems that once would have been formidable or impossible.

The basic breakthrough discussed in Chapter 3 is to extend to realistically complex decisions the key decision and risk analysis principles of (i) predicting probable consequences of alternative actions (using stochastic simulation-based risk models of the relation between actions and their probable consequences, which may include complex, nonlinear interactions and random transitions) and (ii) finding the "best" feasible actions, defined as those that create preferred probability distributions for consequences. This is accomplished via *simulation-optimization models*, in which computer simulation models are used to represent the probable behavior of the system or situation of interest in response to different choices of controllable inputs. Powerful heuristic optimization methods, extensively developed over the past two decades, then search for the combination of controllable inputs that produces the most desirable (simulated) distribution of outcomes over time. Modern simulation-optimization technology substantially extends the practical reach of decision analysis concepts by making them applicable to realistically complex problems, provided that there is enough knowledge to develop a useful simulation model.

Breakthrough 3: Decision-Making with Unknown Risk Models

Simulation-optimization technology provides a breakthrough for deciding what to do if the causal relation between alternative feasible actions and their probable consequences—however complex, nonlinear, dynamic, and probabilistic it may be—is understood well enough to be described by a risk model that can be simulated on a computer. But suppose that the relation between controllable inputs and valued outputs is unknown, or is so uncertain that it cannot usefully be simulated. Then different methods are needed. One possibility, discussed in Chapters 4 and 7, is to *learn from experience, by intelligent trial and error.* In many applications, one might dispense with models altogether, and experiment and adaptively optimize interactions with the real system of interest, in effect replacing a simulation model with reality. (This is most practical when the costs of trial and error are relatively small.) For example, a marketing campaign manager might try different combinations of messages and media, study the results, and attempt to learn what combinations are most effective for which customers. Or, in the domain of healthcare, a hospital might try different drugs, treatment options, or procedures (none of which is known to be worse than the others) with patients, collect data on the results, and learn what works best for whom—the basic idea behind clinical trials. Chapter 4 considers optimal learning and anticipatory decision-making, in which part of the value of a possible current decision is driven by the information that it may reveal, and the capacity of such information to improve future decisions (the "value-of-information" (VOI) concept from traditional decision analysis). Recognizing that any current "best" estimated model may be wrong, and that future information may lead to changes in current beliefs about the best models, and hence to changes in future decision rules, can help to improve current decisions. Chapter 7 also discusses "low-regret" decision-making, in which probabilities of selecting different actions, models to act on, or decision rules are adaptively adjusted based on their empirical performance. Such adaptive learning leads in many settings to quick convergence to approximately optimal decision rules.

A second approach to decision-making with initially unknown risk models is possible if plentiful historical data on inputs and outputs are available. This is to estimate a relevant causal risk model from the available historical data. The estimated model can then be used to optimize decisions (e.g., selection of controllable inputs, or design of policies or decision rules for selecting future inputs dynamically, based on future observations as they become available). Chapter 6 briefly surveys methods of causal analysis and modeling useful for constructing risk models from historical (observational) data; for testing causal hypotheses about the extent to which controllable inputs (e.g., exposures) actually cause valued outputs (e.g., changes in health risks); and for estimating the causal impacts of historical interventions on outcomes. Chapter 7 discusses what to do when more than one possible model fits available data approximately equally well, making it impossible to confidently identify a unique risk model from the data. In this case, model ensemble methods, which combine results from multiple plausible models, can give better average performance than any single model.

Finally, a third possible approach to decision-making with unknown or highly uncertain models, also discussed in Chapter 7, is to seek "robust" decisions—that is, decisions that will produce desirable consequences no matter how model uncertainties are eventually resolved.

Of course, such a robust decision may not always exist. But a rich theory of robust optimization (and of its relations to related innovative concepts, including coherent risk measures and to older techniques such as stochastic programming) has been developed relatively recently, and this theory shows that many risk management decision problems of practical interest can be formulated and solved using robust optimization techniques. Chapter 7 and its references discuss these recent developments further.

Taken together, these relatively recent techniques for dealing with model uncertainty in decision-making constitute a distinct improvement over earlier methods that required decision-makers to, in effect, specify a single best-estimate model (typically based on subjective probabilities). Allowing for model uncertainty leads to new and useful principles for adaptive learning from data and for low-regret and robust optimization. These hold great promise for a wide variety of practical risk management decision problems, as illustrated by examples in Chapters 4 and 7.

Breakthrough 4: Practical Elicitation and Structuring of Probabilities and Multiattribute Utilities

The first three breakthroughs—behavioral decision and game theory, simulation-optimization, and methods for learning risk models from data and, in the interim, for making decisions with unknown or highly uncertain risk models—represent substantial enhancements to or departures from traditional SEU-based decision analysis. Breakthrough 4 consists of methods for making SEU theory more applicable to complex and difficult real-world problems by eliciting preferences and beliefs via techniques that impose less cognitive load on users and/or that achieve greater consistency and reliability of results than older methods. Chapter 5 considers state-of-the-art methods for developing multiattribute utility functions. This is a potentially painstaking task that once involved estimating multiple trade-off parameters and verifying quite abstract independence conditions for effects of changes in attribute levels on preferences. Chapter 5 presents much simpler and more robust methods, developed largely in marketing science, to enable relatively quick and easy development of multiattribute utility functions from simple preference orderings.

Breakthrough 5: Important Real-World Applications

The final category of breakthroughs consists of applications of decision and risk analysis principles to important and difficult fields that have historically relied on other methods. Chapters 5, 9, and 10 discuss applications of expert elicitation, game theory, decision tree analysis, Bayesian networks (for text mining of natural language), and machine learning techniques to the challenges of modeling and defending against adversarial actions. Chapter 8 illustrates the application of multiattribute utility theory to medical decision-making problems, at both the individual and the societal levels, by assessing utility functions for making trade-offs between consumption of sugar-sweetened beverages and risks of morbidity (type 2 diabetes) and early mortality. Chapter 9 discusses vulnerability, resilience, and defense of complex systems—specifically, electric power networks—and compares the insights gleaned from game-theoretic models and considerations to those from less sophisticated methods, concluding that the more sophisticated methods are often very worthwhile. Chapter 6 suggests that many public health decisions that are based on attempts to interpret associations causally would be much better served by applying more objective (and now readily available) methods of causal analysis and risk modeling.

In each of these application areas, and countless others, decision support methods have long been used that do not incorporate the precepts of SEU theory, modern decision analysis, simulation, optimization, optimal learning, or analysis and deliberation using causal risk models of causal relations. In each of these areas, adopting improved methods can potentially achieve dramatic objective improvements in average outcomes (and in entire probability distributions of outcomes). This point is made and illustrated by dozens of examples and case studies in the chapters that follow. Adopting the methods discussed in this book, implementing them carefully, and monitoring and learning from the results can yield breakthrough improvements in areas including marketing, regulation, public health, healthcare and disease risk management, infrastructure resilience improvement, network engineering, and homeland security. The conceptual and methodological breakthroughs presented in the following chapters were selected because they are ready for practical use and because they have been found to create great benefits in practice. Opportunities to apply them more widely are many, and the likely rewards for doing so are great.

CHAPTER 2

The Ways We Decide: Reconciling Hearts and Minds

Edward C. Rosenthal
Department of Marketing and Supply Chain Management, Fox School of Business, Temple University, Philadelphia, PA, USA

The story of Odysseus and his encounter with the Sirens is one of the oldest and most psychologically captivating tales in the Western literary tradition. The sea god Poseidon, you may remember, cursed Odysseus to wander the sea for 10 years, and he and his crew faced countless and remarkable ordeals—among them, getting past the Sirens. The Sirens were known for their beautiful singing; in fact, all sailors who came within earshot of them would compulsively steer their boat closer, only to perish on the rocky shore.

Knowing of the Sirens' reputation, Odysseus was determined to avoid disaster, but he desperately wanted to hear the songs for himself. His solution was to have his men plug up their ears with wax and tie him to the mast, under strict orders not to release him until they were well out of danger. Spellbound by the beautiful voices, Odysseus vehemently insisted that the crew untie him, but his men obeyed the plan and freed him only after they had sailed away.

Breakthroughs in Decision Science and Risk Analysis, First Edition.
Edited by Louis Anthony Cox, Jr.
© 2015 John Wiley & Sons, Inc. Published 2015 by John Wiley & Sons, Inc.

What does this mean for us? Homer's tale is the first known account of someone who, in a cold and unemotional state, was able to anticipate his urges when in an aroused state and take action to prevent those urges from being acted upon, thus avoiding a disastrous outcome. Even thousands of years ago, we realize, people recognized the very real conflict between the emotional and unemotional, or the hot and the cold, facets of our decision making. However, even today, the coexistence and interplay between these very different drivers of our behavior are not as well understood as they could be.

Decision theorists, as well as psychologists, are eager to answer the question, How do we make decisions? This chapter explores the two different sources to which we typically attribute our decisions when challenged to explain them—hearts and minds—in terms of their roles, their interactions, and the conflicts between them. First, we will explore some biological and psychological underpinnings: Do we actually "make" decisions, or are our choices and actions determined for us by unconscious deterministic processes? What are the Systems 1 and 2 that psychologists often mention? How might evolution have shaped our decision processes?

Next, we will look at scientific attempts to provide a sound foundation for decision making. We'll peek at the underpinnings of subjective expected utility theory (SEU)—a powerful model meant to formalize our preferences among different actions when their outcomes are uncertain. After that, in order to understand strategic behavior, we will quickly review some important results from theoretical and empirical game theory. Game theory and SEU together provide a prescriptive methodology for how to act in well-defined situations with the following features: different outcomes (or consequences) of choices are possible; the different outcomes are not all equally desirable (since otherwise it would not matter what one did); the outcomes are uncertain (and their probabilities depend on what actions are taken); there may be multiple decision makers whose choices are jointly responsible for bringing these different outcomes about, or at least affecting their probabilities; and the preferences of different decision makers might conflict. For the remainder of this chapter, I use the word "rational," to refer to decisions and behaviors that would be suggested by an SEU or game-theoretic model of a particular situation.

Game theory, together with SEU, forms an intimidating edifice. This normative approach represents the cold and unemotional side of life. And since SEU and game theory are scientifically formulated to provide the best responses and, loosely speaking, best overall outcome probabilities when faced with a decision, it seems reasonable to think that no other actions can be superior to the rational ones prescribed by the theory. But, after having seen the highlights of these approaches, we will explore their limitations—the chinks in the armor, so to speak. How do our emotions enter into our decision making? How should they? Can emotionally charged decisions work out better than the rational ones? To explore these questions, we will review *prospect theory*. This descriptive theory shows that the normative framework of SEU does not seem to accurately describe or explain actual human behavior, and identifies and describes systematic differences between actual and SEU-consistent decisions. We then go beyond prospect theory to observe other important examples of how *homo sapiens* seem to interpret their world and act on certain systematic biases. These examples include recent findings in the fields of neuroeconomics, behavioral economics, experimental game theory, moral decision-making, intertemporal choice, and consumer behavior.

While prospect theory and behavioral decision theory describe our preferences in certain situations, they do not address how we do or should choose in situations of *conflict* or *cooperation*. Behavioral game theory, on the other hand, contrasts the normative solutions of game theory with the choices that people really make when their outcomes depend on others' actions too.

Finally, the chapter wraps up with a discussion of where we stand, where to go from here, and how to capitalize on our newfound perspective.

DO WE DECIDE?

How do we decide to do something? What compels us to act? Clearly, these questions lurk within any discussion of decision making, but to address them is to get caught up in the ancient philosophical topic of free will versus determinism. Spinoza (1677) likened our thought processes to the flights of projectiles. Our trajectories are determined, and to believe that we choose them is no different than proposing that the projectiles' thoughts (if they had any) could alter their arcs. For William James (1884), we react to a stimulus physiologically; our *subsequent* mental perception of our bodily state is registered as an emotion. Our heart doesn't pound because we are afraid; we are afraid because our heart pounds. Even Sartre (1943), who put so much emphasis on our choosing responsibly, described how our conscious thoughts come only after the bets have already been made, so to speak. To put it differently, when we're picking which track to take, the train has already left the station.

This line of reasoning has been updated by Daniel Wegner (2002), who cites research such as Libet (1981) and Libet, Wright, and Gleason (1983) to assert the same thing: the possession of a conscious will that brings about our actions is an illusion. Libet's research involved timing when a person is first aware of willing an action (such as moving a finger), and when the action takes place (when the finger actually moves). Of course, as you would suspect, willing the finger to move precedes its movement. However—and this is the crucial finding—more than 300 milliseconds *prior* to the conscious willing of the action, there is a so-called "readiness potential," that is electrical activity in the brain, that seems to initiate the process, although no one knows how. The neurological correlates of choice precede conscious awareness.

While this is not the forum for an extensive discussion of free will—what is it, how might it exist (if at all), and how free can it really be?—it is important to realize that sometimes we make a decision to do something, and *we don't understand why*. (Conversely, as demonstrated in split-brain experiments, the reasons that we give ourselves for our choices may have much to do with narrative plausibility and little or nothing to do with the real drivers of choice: we often rationalize our actions after the fact.) With certain decisions, such as which move to make in a chess game, the resulting action certainly seems to be the result of purposeful (and, we would like to think, rational) cogitation. But as we discover that so many of our decisions seem to occur without conscious deliberative thought, we need to remember that those decisions might be—whether or not we want to admit it—already determined outside our conscious deliberation. Even when formal decision analysis (such as SEU theory) is applied to make a decision, so that the choice process is explicit, formation of beliefs and of preferences for outcomes may be shaped by subconscious processes. This has been demonstrated by psychologists (Paul Slovic and colleagues), who have shown that beliefs about outcome probabilities and preferences for outcomes are correlated: both tend to be shaded to make choices that are perceived as "good" (positive affect) appear to be more likely to produce preferred outcomes than choices perceived as "bad" (negative affect). The *affect heuristic* shows that attitudes toward choices affect perceptions of their probable consequences, rather than the other way around.

BIOLOGY AND ADAPTATION

Let's extend this discussion further, while steering clear of the black hole of armchair philosophy. What do psychologists tell us about our decision making? Is it true that some decisions are arrived at slowly and thoughtfully, while others are made quickly and automatically? The answer is yes, and Kahneman (2011) has organized decades of research on this important dichotomy, which crystallized around the turn of this century. The two different decision engines are often referred to as System 1 (quick and automatic) and System 2 (deliberate and slow), as coined by Stanovich and West (2000). Be clear, however, that this "system" terminology is just a convenient shorthand. There are no actual such systems (let alone such an apparatus!) in our brains, and there is no one area of the brain that is "owned" by either one. To borrow from Kahneman, some examples of tasks we solve using System 1 are detecting that one object is more distant than another, expressing one's disgust, and understanding a simple sentence. System 2 takes care of things like searching your memory to identify a sound, or checking the validity of a complex logical argument. While you can see from the examples that System 1 and System 2 do not exactly hew to the distinction between emotional and rational that we are interested in, there is certainly a close correspondence.

Given that the free will versus determinism issue invokes biology, and that the origins of the System 1 versus System 2 duality are ultimately rooted in multiple lobes of our brain, it is natural to speculate as to how our brain function and our resulting preferences and behaviors might have evolved over time. It is common place for authors to state arguments appealing to our having developed in the Pleistocene epoch: for example, that men prefer women of certain shapes and ages because these observable traits are correlated with fertility, or that people prefer foods with high sugar and fat content because these foods supply both short- and long-term energy. Interesting discussions of these preferences emerge because some of our hardwired inclinations, albeit useful in the Pleistocene, are ill-adapted nowadays; for example, eating foods rich in sugar and fat may increase our risk of diabetes, cancer, and heart disease.

Evolutionary biologists speak of *fitness*, which is the relative ability of an individual to survive and pass on its genes. The standard approach to evolution and behavior is this: when mutations occur that change our behavior, the altered behavior patterns are put to the test in the wild. If the new behaviors prove to be adaptive—that is, they endow their owner with at least as much fitness as the previous set of behaviors—then they obtain a foothold in the population. If not, they will fade away from one generation to the next, if not immediately. Thus, we tend to think of evolution as a lengthy process that unforgivably weeds out subpar performers.

Given that I have advertised SEU and game theory as prescribing the "best" actions to take in well-defined decision problems, and given that thousands of generations of evolutionary honing would also seem to result in highly effective (although perhaps not "best") behaviors, let's study a situation with two possible actions—an emotional and a rational choice—and compare their consequences. Frank (1988) pioneered the idea that people's emotional makeup might have evolved to play an important strategic role in our social interactions. To follow his argument, let's say that you have been wronged by someone. They've spread false rumors about you on social media and you are pretty angry about it—so angry, in fact, that you might exact some major and disproportionate revenge. On the other hand, your friends advise you to let the whole thing blow over and by tomorrow people will have half-forgotten, having already moved on to the next drama. They point out that if you retaliate, it will backfire and make you look even worse.

So what are you going to do? Your friends have already crunched the numbers, so to speak. Their cost-benefit analysis has you looking worse by reacting than if you let it all slide. The passive response is the rational response. But on the other hand, you're hell-bent on getting revenge!

And what if your lust for revenge is unstoppable: you will heavily retaliate, despite the fact that it is not in your best interest, and now both of you will have suffered. But here comes the point: if you are consistently *committed* to acting on your emotions, and your nemesis had been aware of your hotheadedness, he would not have wronged you in the first place!

Frank's idea is that sometimes, seemingly irrational behavior (like acting on your anger), stems from emotional predispositions that, although contrary to what seems to be our self-interest, can work out for the better. This *commitment principle*, though, leads us to an apparent paradox: how can irrational behavior become superrational, that is, out-perform rational behavior?

The key here, and in the analysis of other so-called commitment devices in the literature, is that the conflict between the rational model—in which you are acting in apparent self-interest—and the commitment model stems from an apples-and-oranges misunderstanding of the time horizon involved. When your friends advised you not to retaliate, their reasoning was that an aggressive response would make you worse off. Your friends were right—but they were thinking short term. The commitment strategy, on the other hand, and the emotional predisposition that one needs to help implement it is adaptive in the long run.

Love—by which I mean a deep and lasting emotional attachment—is another example where a commitment strategy may lead to inferior short-term outcomes but can triumph in the long run. Couples that do not love one another would break up when the going gets rough; but love provides a way to bind people together for the long term. The romantic view (which is, of course, arguable) is that the long-term partnership will provide greater happiness for each—as opposed to, for example, a string of briefer relationships.

One last point that needs to be made on evolution and emotional makeup is that not all longstanding behaviors (emotionally based or otherwise) are necessarily optimal ones. Just because certain behaviors have survived the years does not mean that they were ever optimal, or even forged through ruthless adaptation in the wilds of the Pleistocene. Plenty of human characteristics have stuck around for generations—like being moody, having a memory like a sieve, or a predilection for pyromania. I just want to make sure that you do not feel the urge to craft a fitting story to explain every quirky character trait in the human arsenal.

SEU AND GAME THEORY

Before we get any deeper into the ways in which emotional and rational decision-making can clash, we need to better understand exactly the foundations of rational decision making (by which I mean SEU and game theory). Expected utility theory has its origins in a work of Daniel Bernoulli (1738), in which he introduced the notion of "utility." Bernoulli's motivation was to find a solution to the so-called St Petersburg problem. In this game, a fair coin is flipped until the first heads appears. This event happens, say, on the nth flip and the person playing the game is paid 2^n dollars. The player, however, must pay a fee to play. How much is the game worth?

Since the probability of the game ending on the nth flip is $1/2^n$, the expected value of the game is

$$\frac{1}{2} \times 2 + \frac{1}{4} \times 4 + \frac{1}{8} \times 8 + \cdots = 1 + 1 + 1 + \cdots = \infty.$$

The paradox is that although the game has an infinite expected value (meaning in effect that people ought to be willing to pay a huge amount to play the game), no one would pay a significant sum here. The key, Bernoulli realized, is that the actual worth of the money to us is not proportional to the monetary amount. Ten million dollars, for example, is worth a lot more to you than one million dollars—but not 10 times as much.

Bernoulli used a logarithmic function to represent the value, or utility, of a specific amount of money. If we represent the utility of a sum x of money as $U(x) = k \log(x)$, for example, then the infinite sequence for the value of the St Petersburg game will converge to a finite and reasonable number.

About 200 years later, von Neumann and Morgenstern (1944) revisited this problem, with the goal of developing a more comprehensive approach to utility and its role in decision making when presented with uncertain (i.e., *risky*) outcomes. Let X represent the set of all possible outcomes in a particular situation. Von Neumann and Morgenstern's first requirement was that there is a *complete* preference relation on X, that is, for any two outcomes A and B in X, either A is preferred to B or B is preferred to A (preference here is weak preference, where ties are allowed).

Next, the set of preferences on X must be *transitive*, that is, for any three alternatives A, B, and C, if A is preferred to B and B is preferred to C, then A must be preferred to C. At this point, let us define a *lottery* over X as a *probability distribution* on X (an assignment of probabilities to the outcomes in X such that the probabilities are nonnegative and add to one). Let L be the set of all possible lotteries on X. Let $\gtrsim$ represent the preference relation on L as follows: if A is preferred to B, we write $A \gtrsim B$. Let $\approx$ denote indifference, that is, if $A \gtrsim B$ and $B \gtrsim A$, then $A \approx B$, meaning they are equivalent. Given that $\gtrsim$ must be complete and transitive, the following are the additional von Neumann–Morgenstern conditions for $\gtrsim$:

Relation $\gtrsim$ satisfies the "sure-thing principle": for lotteries G, F, g, and f such that $G \gtrsim F$ and $g \gtrsim f$, and any probability p, then $p \times G + (1 - p) \times g \gtrsim p \times F + (1 - p) \times f$. The sure-thing principle says that for any weighted average between G and g, and the same weighted average between F and f, you would prefer the G/g scenario to the F/f scenario.

Relation $\gtrsim$ satisfies continuity (or comparability), which is stated as follows: Given three lotteries F, G, and H such that $F \gtrsim G \gtrsim H$, then there exists probability p such that $p \times F + (1 - p) \times H \approx G$. In other words, this condition states that there must be some way to average F and H so as to be indifferent to G.

Now, von Neumann and Morgenstern simply quantified the preference relation as follows: we assign a utility function $U(x)$ on the real numbers such that $U(A) \geq U(B)$ if and only if $A \gtrsim B$. In what is called the SEU model, as stated by Savage (1954), we specify that the decision criterion is expected utility maximization, although the probabilities used by any one decision maker are subjective. A rational decision-maker is one who will make decisions according to the above conditions on their preferences among a set of risky alternatives. In other words, a rational decision-maker will choose the lottery, among those available, that maximizes her expected utility.

One difficulty with the expected utility approach has been in the subjective elicitation of utilities from decision makers, as discussed further in Chapters 5 and 8. Different methods have been proposed over the years to overcome this concern, but these methods can lead to different results. Some methods require decision makers to have a precise knowledge of a probability distribution, and misperceptions in such estimation can lead to a violation of the expected utility conditions. To escape these difficulties, Wakker and Deneffe (1996) proposed a "gamble-tradeoff" approach, in which utilities can be surmised while bypassing the need to pin down exact probabilities.

The expected utility model treats preferences among uncertain outcomes, but it does not inform us how to make a decision when the outcomes depend on the decisions of others. Such situations require

(a)

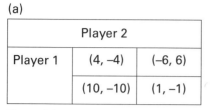

	Player 2	
Player 1	(4, –4)	(–6, 6)
	(10, –10)	(1, –1)

(b)

	Player 2	
Player 1	(1, –1)	(–1, 1)
	(–1, 1)	(1, –1)

Figure 2.1 Zero-sum games.

strategies, and this is where game theory needs to be called up to the front lines.

A *game* is a situation with multiple decision-makers, each one of whom is called a *player*. Each player has a certain set of actions, or *strategies*, to choose among. The joint result, when each of the players selects a strategy, is called the *outcome*. These outcomes are usually measured numerically and are often referred to as *payoffs*. Like SEU, game theory is a normative approach to problem solving; it prescribes how decision-makers ought to act.

The simplest game to describe is a 2×2 game. Here, there are two players, each of whom has two strategies from which to select. One way, we will depict such games is in so-called *strategic form*, where a *payoff table* is used to enumerate the strategies and the payoffs. Another way to describe games is called *extensive form*, where a tree diagram is used to indicate the temporal sequence of moves by the players.

In some games, called *zero-sum games*, one player loses what the other one gains. Such games are strictly competitive. In nonzero-sum games, though, the payoffs to the players are not necessarily opposed and need not be related.

Before describing solutions for games, I want to give some brief examples of games that will be useful to us. In the 2×2 games in Figure 2.1, Player 1 picks among the rows in the payoff table and Player 2, making a simultaneous decision, picks among the columns. The outcomes are listed as utilities, or perhaps monetary amounts, issued to the two players, respectively.

In Figure 2.1a, Player 1 controls the rows. He observes that choosing the second row always leads to a better payoff for him than choosing the first row. For example, if Player 2 were to pick the first column, Player 1's payoff of 10 in row 2 is superior to the payoff of 4 in row 1. Likewise, if Player 2 were to choose the second column, Player 1 is better off in row 2 with a payoff of 1 as opposed to row 1 with a payoff of −6. We say that row 2 *dominates* row 1 for Player 1.

Similarly, for Player 2, column 2 dominates column 1. Thus, the best strategy for Player 1 is to choose row 2 and the best strategy for Player 2 is to pick column 2. This pair of strategies leaves the players with the outcome (1, −1). While this is not a good outcome for Player 2, it is the best outcome that she could achieve.

This thought process introduces the notion of an *equilibrium*, the most important solution concept in game theory. Given that that Player 1 is going to play row 2, Player 2 cannot improve her payoff by switching to column 1. And, given that Player 2 is going to play column 2, Player 1 cannot improve his payoff by switching to row 1. Since neither player can improve their outcomes by switching from their dominant strategies, we say that the strategy pair row 2, column 2, with outcome (1, −1) is an equilibrium.

Now consider the game in Figure 2.1b. Neither row 1 nor row 2 is a dominant strategy for Player 1, and neither column dominates for Player 2 either. It turns out that in this game, the players will have to resort to *mixed strategies*, as opposed to the single, or *pure* strategies, to reach an equilibrium. (In mixed strategies, the players employ a probability distribution over their strategy set.) The equilibrium strategy pair in this case is (1/2, 1/2) for both players, and the way that the each player would need to implement the mixed strategy is to randomize: to use a random number generator or other such device that will select between their two strategies with probability 1/2 each.

Zero-sum games, while useful, are not appropriate to model the complexity of most strategic interaction. Generalizing the model to include outcome pairs that are not diametrically opposed opens up a lot of possibilities for exploration and analysis. Probably the most well-known 2×2 nonzero-sum game is the Prisoner's Dilemma, or PD. In this game, each player has the choice either to cooperate with the other player, or to defect. Mutual cooperation is a good outcome for both players. However, as we will see in Figure 2.2, the way the game is structured, defection is a dominant strategy.

In the PD in Figure 2.2, defection is the dominant strategy for both players. Unfortunately, this strategy pair leads to a unique equilibrium payoff of (1, 1), which leaves both players a lot worse off than the outcome of (3, 3) had they both cooperated. For this reason the PD is a maddening game; what's more, it seems to be an

	Player 2	
Player 1	Cooperate	Defect
Cooperate	(3, 3)	(0, 5)
Defect	(5, 0)	(1, 1)

Figure 2.2   The Prisoner's Dilemma.

appropriate model for numerous situations in social and commercial interaction (Rapoport and Chammah, 1965; Parkhe, Rosenthal, and Chandran, 1993; Mudambi, 1996).

One might ask at this point, do all games have equilibrium solutions? Fortunately (because otherwise we would be stumped as to how to play games!), the answer is yes. Von Neumann (1928) proved that there always exists an equilibrium in any zero-sum game, found from using *minimax* strategies, and Nash (1950) proved a similar result for nonzero-sum games. Intriguingly, however, many nonzero-sum games have more than one Nash equilibrium. This is somewhat unfortunate, because the different equilibria may have very different outcomes; some of them would be difficult to implement in a real-life scenario; and players may have difficulty figuring out which one to go for.

Game theory has grown far beyond the fundamental results that I have sketched here. Myerson (1991), Dutta (1999), and Binmore (2007) are three texts that provide in-depth expositions of the subject. One topic of great importance is how to create and solve games in situations of asymmetric information, that is, when one player is much better informed than the other. Another sphere of interest is found in cooperative games, in which researchers are interested in developing fair ways of apportioning costs or benefits when players join together to improve their payoffs. And mechanism design forms another significant frontier of study, where research involves a reverse-engineering of sorts: first identifying a desirable outcome and then developing rules of play to incentivize the players to steer in that direction. While these different directions in game theory are important for political and business applications, they aren't necessary for our purposes. In fact, with what we now know about the rational approach to decision-making—SEU and basic game theory—we can more fully grasp what lies ahead: further discovery of how, and the nascent understanding of why, human decision-making diverges from rational prescriptions.

PROSPECT THEORY

Prospect theory, which was introduced by Kahneman and Tversky (1979), points to a variety of ways in which the preferences of most people do not follow the prescriptions of SEU. But well before prospect theory, Allais (1953) had constructed a numerical example illustrating that most subjects' preferences are inconsistent with von Neumann–Morgenstern utility theory. To see this, consider the following four lotteries in Figure 2.3:

Allais found that most people prefer lottery J to lottery K, because J promises one million dollars for sure; but that most people also prefer lottery M to lottery L. However, for a person to exhibit such a set of preferences is to contradict SEU. While troublesome, the Allais paradox was probably less worrying to proponents of SEU than it would appear, perhaps because SEU advocates could claim, for example, that people answered the questions inconsistently on account of their not understanding their (own) preferences. The arrival of prospect theory, however, was a game-changer.

Unlike SEU, prospect theory is supported by a wealth of experimental evidence. It is based on three main pillars:

1. "Prospects" (typically, monetary values) are evaluated relative to a starting reference point.
2. A cognitive principle of diminishing returns (or "sensitivity") exists.
3. Losses loom larger than gains.

J:	$0	1M	5M
	0	1	0

K:	$0	1M	5M
	0.01	0.89	0.1

L:	$0	1M	5M
	0.89	0.11	0

M:	$0	1M	5M
	0.9	0	0.1

Figure 2.3   The Allais paradox. Each outcome has an associated probability.

Two other features are also very important. These are as follows:

1. People are risk-taking for losses but risk averse for gains.
2. People misweight probabilities.

Let's briefly study each of these points.

Working with our restriction to measuring monetary outcomes, expected utility theory, it turns out, goes along with Bernoulli's view that utility is measured by one's wealth. But a key finding of Kahneman and Tversky is that we need to measure *changes* in wealth, not absolute wealth. Since SEU does not take into account a baseline, or starting point, for making a decision, the prescriptions given by SEU will not correspond to how people make decisions. Kahneman (2011) gives the following example as evidence: Suppose Anthony currently has one million dollars while Betty has four million dollars. Each of them is offered the following choice: to own two million for sure or else to gamble on a lottery in which there is an equal chance to end up with either one million or four million.

The expected value of the gamble is 2.5 million. Anthony, if he is like most people, would prefer to keep two million for sure. Betty, however, would be very unhappy to take the sure thing and thus reduce her wealth from four to two million; if Betty is like most people, she would prefer the lottery. Expected utility theory offers no insight regarding their present states; it posits that Anthony and Betty would both make the same decision.

The phenomenon of diminishing returns is common to both SEU and prospect theory, so we won't dwell on it except to ask, which would give you more pleasure: increasing your wealth from zero to one million dollars, or increasing your wealth from 56 to 57 million dollars? (Most of us would be ecstatic in the first instance but only mildly charged in the second one.)

When we say that "losses loom larger than gains" (Kahneman, 2011), we realize that SEU was not formulated to consider possible *losses* in wealth. Introducing the prospect of losses is not only important in its own right, but it also exposes a human quirk that is inconsistent with SEU. Would you accept a lottery with 0.5 chance of gaining $125 and a 0.5 chance of losing $100? Probably not, despite the positive expected value. Empirical work shows, incidentally, that for most people, gambles like the one I just offered you would come out indifferent if the gain were about double the loss (in this case, in the neighborhood of $200). In other words, for most people, the pain of loss is roughly double the pleasure of winning (at least, for a wide range of monetary quantities).

Another finding of prospect theory is that most people's preferences when presented with lotteries over losses differ from those for gains. For example, consider an even chance of winning either $100 or $0. If you are like most people, you would prefer a sure $45 to the gamble (whose expected value is $50). Such risk-averseness (represented by a concave utility function) is consistent with SEU and the diminishing returns property we already discussed. But now consider the realm of losses: suppose you faced a lottery with an even chance of losing $100 or $0 as opposed to a sure loss of $45? Faced with these sorts of situations, most people become risk-seeking, that is, will prefer the gamble. This surprising result does not jibe with SEU, since the utility function here is convex in the loss domain.

Finally, what does it mean to "misweight" probabilities? Kahneman and Tversky discovered that their experimental subjects would seem to make certain mental adjustments or translations of probabilities presented to them. In general, most people underestimate most probabilities; in other words, when presented with a probability, say, of 0.40 for a certain event, the weight that most people will mentally assign works out to less than 0.40. However, most of us also overestimate very small probabilities. Most of us, when told of a one-in-a-thousand chance, will mentally assign a figure greater than 0.001. Here are a couple of examples: when offered a choice between a 40% chance to gain $5000 and an 80% chance to gain $2500, most people prefer the 80% chance of $2500; when one mentally adjusts both probabilities, the 80% offer retains more of its value.

But how do people react to very small probabilities? To understand this, consider a 0.001 chance to gain $5000 and a 0.002 chance to gain $2500. Given this choice, most people prefer the 0.001 chance at $5000. It would appear that they overweight the 0.001 more than the 0.002; the result is that the two probabilities look almost the same, and then the $5000 is more attractive.

The realization that people misweight probabilities spurred researchers to develop a theory that could reliably model this behavior. The idea here is to transform probabilities into the somewhat idiosyncratic weights (the over- and underestimates) that people seem to apply in their decision making. Such a transformation would have to be non-additive, and various studies presented ways to calculate expected utilities with respect to the transformed scale (Quiggin, 1982; Yaari, 1987; Schmeidler, 1989). These efforts were incorporated into a more general form of prospect theory called cumulative prospect theory (Tversky and Kahneman, 1992; Wakker and Tversky, 1993). In cumulative prospect theory, rather than treating individual outcomes with their singleton attendant probabilities, the outcomes are rank ordered and cumulative probabilities are associated with nested subsets of outcomes. While it was shown that cumulative prospect theory accounts for the basic elements of prospect theory, it nevertheless remains a formalization of the theory, with few additional behavioral insights to be gleaned from it.

BEHAVIORAL DECISION THEORY

Beyond prospect theory, there are a number of other well-documented biases that people exhibit. Let us use the term *behavioral decision theory* to encompass, quite simply, the study and categorization of how people actually make decisions. As such, behavioral decision theory would include prospect theory in addition to the rather *ad hoc* biases that I will describe below (as well as others that we don't have the space to include).

There are two important and surprising ways in which people can be strongly influenced prior to their making a decision. Psychologists have studied *priming* in a number of contexts. In a notable study (Shih, Pittinsky, and Ambady, 1999), Asian-American women were administered a mathematics test. Prior to taking the test, the women were separated into three groups: one group received a questionnaire with items related to their ancestry (designed to focus the women on their Asian heritage); the second group's questionnaire emphasized women's issues (to focus those participants on their gender); and the third group answered neutral questions. The idea of the study was to see if "priming" the women with cues to their Asian background would improve their test scores (consistent with the cultural stereotype that Asians have superior math skills), or if cues invoking their feminine side would lower their test scores (consistent with the stereotype that women are inferior at math). Remarkably, the performances on the exam followed the stereotyped cultural pattern: the highest scoring group was the ethnically primed subset; the lowest scoring group was the gender-primed group, while the neutral group was intermediate to the other two.

While priming largely predisposes people to a certain state of mind, *anchoring* is a technique where the prior information establishes a baseline bias with respect to particular responses. Split a group of people into two: ask people in the first group if the population of Buffalo is more or less than 500,000; ask those in the second group if the population of Buffalo is more or less than 100,000, and then ask people in both groups to estimate Buffalo's population. The most likely result is that the responses from the first group will be much higher than those from the second group.

Anchoring can take effect in much more subtle ways. Ariely, Loewenstein, and Prelec (2003) performed an auction of items such as wine, chocolates, and cordless keyboards. But just prior to the auction, they asked the participants to write down the last two digits of their Social Security numbers, as if these two-digit numbers would represent a hypothetical auction bid. When the auction was over, the participants dismissed the notion that their writing down an irrelevant number could possibly influence their bids for wine and chocolate. But the results showed otherwise: there was a clear correlation between those "random" two-digit numbers and the subjects' actual auction bids. Those writing down numbers from 80 to 99 bid, on average, from 2 to 4 times as much as those writing down numbers from 00 to 19.

Another important bias in behavioral decision theory is described as the *availability heuristic*. Here, when people are attempting to estimate the likelihood of something, if they can think of a readily available instance (say, a stereotype), the identification they make and its familiarity will cause them to overestimate the original category. The most commonly cited example involves terrorism. When terrorism is in the news, people may avoid flying overseas; instead, they may opt for an auto trip on Interstate 95, which is statistically more dangerous than a transoceanic flight.

The *representativeness heuristic* is a bias not too distinct from the availability trap. With this behavior, people are trying to categorize: does a particular item or subject belong to a certain group? The mistake people often make is to substitute a certain stereotype in appraising the categorization, and this throws off their judgment. Tversky and Kahneman (1983) gave the famous example of Linda, who is thirty-one years old, single, outspoken, and very bright. She majored in philosophy. As a student, she was deeply concerned with issues of discrimination and social justice, and also participated in antinuclear demonstrations.

Is Linda more likely to be a bank teller, or a bank teller who is active in the feminist movement? Most people believe Linda is more likely to be a bank teller active in the feminist movement. However, this choice defies simple logic: since the category of bank tellers includes those who are feminists, and is therefore a larger category, the probability of Linda being a bank teller must be larger than the probability she is a bank teller active in the feminist movement. The trap, more generally, with the representativeness heuristic, is that very often, people observe behavior—generally with small sample sizes—and from the patterns they detect, they make false generalizations.

Another bias that hinges on the associative properties of our memories is known as the *confirmation bias*. This is the tendency so many of us have to accept information that is consistent with our beliefs, and to reject statements that are at odds with our beliefs. Two people can watch the same debate and have wildly opposing views on what was said.

There are a number of other well-documented biases in behavioral decision theory, but let's limit ourselves to one more that is particularly relevant to our topic. This last bias is known as the *affect heuristic* (Zajonc, 1980; Finucane et al., 2000). When facing choices, we very often develop quick and automatic likes and dislikes and fears and comforts, and these "gut reactions" end up making the decisions for us. Finucane et al. (2000) examined our evaluations of risk and benefits for items like nuclear power or food preservatives. The results were that the experimenters were able to manipulate the subjects' assessments of risk and benefit by introducing positive and negative affects. In other words, our perceptions of the benefits or risks of such things as nuclear power and food preservatives can be altered by the manipulation of our associated emotions. It is important to note, however, that in this study, the presence of the affect heuristic was much more pronounced when the subjects were under time pressure. In other words, when the subjects had more time to answer the questions, their answers were less affected by the affect manipulation.

The amount of time it takes us to make a decision, in fact, is a key feature, which was underlined in the implicit association test (IAT), introduced by Greenwald, McGhee, and Schwartz (1998). In this test, subjects are asked to make associations among different categories and attributes and respond with a correct matching. For example, flowers and insects might be two different categories: pleasant and unpleasant might be two attributes. Subjects were required to press certain attribute response keys when presented with names from the categories on a computer screen. It turns out that subjects' response times were quicker when associating certain categories and attributes (such as flower and pleasant) than other categories and attributes (such as insect and pleasant).

This result is not surprising. But that's just the warm-up. It turns out that Korean and Japanese subjects differed in their response rates for associations such as (Japanese + pleasant) or (Korean + pleasant). In the United States, white subjects' response rates differed when associating (White + pleasant, which had quicker responses) as opposed to (Black + pleasant, which had slower responses)—and here's the real point—even when the subjects explicitly disavowed any element of racial prejudice. As could be expected, the IAT has been the subject of intense scientific scrutiny (Nosek, Greenwald, and Banaji, 2007) and has also garnered significant attention in the popular press. But as with the affect heuristic, a small amount of extra response time suffices to elicit answers that are consistent with the views that the subjects (explicitly) self-report.

We have seen enough evidence thus far to understand that the differences between System 1 (as we have called it) and System 2 are real and significant. Before Systems 1 and 2 became popular with psychologists, behavioral economist Richard Thaler coined the catchy terms "Econs" and "Humans" to capture the nature of the dichotomy. Econs, of course, make decisions according to the rational program of SEU and game theory, while Humans are imperfect decision-makers whose biases are exactly those we have been describing. Another way to describe our dichotomous nature is to call our two cognitive systems "automatic" and "reflective" (Thaler and Sunstein, 2008). It is important to realize, though, that both of these systems can evolve in their reactions and responses in certain contexts. As Thaler and Sunstein point out, when we first learn to drive, our reflective system slowly puts us through a sequence of behaviors; once we become experienced, it is our automatic system that takes over the controls.

DECISIONS WITH A TIME HORIZON

Now let's turn our attention to other, somewhat different, drivers of our decision processes. The previous research involves asking experimental subjects about their preferences at a certain moment in time, or analyzed respondents' ability to make immediate categorizations, where their response times belie their claims to be unbiased. But harkening back to Odysseus, how well do we align our present desires with our future utility? Research in *intertemporal choice* has begun to give us answers in this domain. Which would you prefer: $15 right now or $20 a month from now? About half of all respondents to this question prefer the $15 right away. But, interestingly, most of *those* respondents would prefer to receive $20 thirteen months from now as compared to $15 in a year. Note the *dynamic inconsistency* of such a preference: if they were asked after a year had elapsed, to now choose between $15 right away or $20 in a month, most would go for the immediate $15—suddenly overruling their previous, more patient, choice.

Thaler (1981) conducted a more comprehensive survey. He asked respondents to imagine they had just won, for example, $15 in a lottery; how much would they have to receive, a month from now, to be equal to the immediate gain of $15? In his study, the immediate prize of $15 was worth the same to people (as measured by the median response) as $20 in a month. As Thaler points out, making people wait for a prize forces them to exert some mental effort. If we measure this effort by using the notion of compound interest, the 33% premium to wait a month works out to an interest rate (compounded daily) of 354%. Thaler also asked, what if they had just won $250? What if they had to wait a year? And, what if they had to *pay* $15 now? What penalty would that equate to in a year's time?

With the larger sum of $250, the median 1-month response was $300; here, the 20% premium works out to an interest rate of 219% (compounded daily). When thinking about waiting a year, the $15 worked out to a (median response) equivalence of $50 (an interest rate of 121%), while the larger $250 was the same as waiting a year to receive $350 (an interest rate of 34%). Even more interesting, people's perceptions changed when contemplating the penalty condition: the $15 penalty paid now was the same, in the median response, as a $20 penalty a *year* from now. In other words, respondents typically did not see much benefit to postponing their penalties.

The specific lessons from this research were threefold: first, as the time to wait grows, people's discount rate decreases, meaning that we are very impatient when waiting short periods of time but become less impatient in the long run. Interestingly, this tendency, called *hyperbolic discounting*, was first documented for a significant minority of pigeons (Ainslie, 1974) and then in humans as well (Ainslie and Herrnstein, 1981). Our discount rates decrease as monetary rewards increase; and, when considering losses, we'd just as soon pay now and be done with it. Research in intertemporal choice has penetrated into areas such as consumption self-control (Wertenbroch, 2003) and drug addiction (Johnson et al., 2010). Researchers have already begun to discover what underlying *neural substrates* (or neural *correlates*) are related to our intertemporal choice behavior. Some early work on the neural substrates of time-dependent choice was carried out by Kable and Glimcher (2007) using functional magnetic resonance imaging (fMRI) technology, while Weber and Huettel (2008) show that the neural correlates of decision making under risk differ from those for intertemporal choice.

MORALS, EMOTIONS, AND CONSUMER BEHAVIOR

One important element of our decision-making that is often ignored in the decision-theoretic literature is the moral dimension. All of us have experienced decisions in which our reflective thought points in one direction, but our moral compass keeps us from traveling down that road. It turns out that what we would call "moral judgment" has traditionally been supposed to have a rational and an emotional component, but research such as Greene et al. (2001) and Greene and Haidt (2002) shows that the emotional component is the stronger one. Consider a runaway train that is about to crush five people to death. The only way to save them is for you to flip a switch to divert the train onto a different track, where it will kill one person. Will you divert the train? Most people would. Now consider the following variation: the same train is still headed toward those five unlucky souls, but the only way to stop it is to push a fellow bystander in front of the train. Would you go ahead and push that person in front of the train? Most people would not.

Ethical philosophers have struggled to come up with a way to justify why our divergent behavior in the two cases is morally justifiable. But Greene et al. (2001) scanned subjects with fMRI technology to show that the thought of pushing that unfortunate individual engages our emotions in a way that merely diverting the train does not. In other words, respondents to this and other dilemmas were scanned while contemplating their decisions, and areas of the brain such as the medial frontal gyrus, posterior cingulate gyrus, and angular gyrus—already known to be associated with emotion—were more active for the very personal task of pushing that individual than for the rather impersonal task of flipping the switch. Haidt (2003) reminds us how moral emotions, such as disgust, support our evolutionary fitness in the same sense of commitment as does anger that we already discussed (Frank, 1988). Disgust plays a more direct role in our survival; its strength as an emotion will, for example, prevents us from eating harmful food (Rozin, Haidt, and McCauley, 2008). And further, our feelings of disgust are very difficult to overturn through reflective thought. Haidt and Hersh (2001) asked subjects to rate the acceptability of a voluntary sexual encounter between a 25-year-old man and his 23-year-old adopted sister who grew up together, given that the sister uses birth control and the man would wear a condom. Neither socially liberal nor socially conservative respondents found this act to be acceptable, although they were generally unable to provide cogent and coherent justifications. In other words, most people were reduced to saying things like, "it just isn't right." Over the past decade, studies like Moll et al. (2002) and Fitzgerald et al. (2004) have started to uncover the complex neural mechanisms underlying disgust.

At this point, we need to make it clear that emotions are necessary to our very functioning, and are not simply evolutionary artifacts. Star Trek's Mr Spock was an emotionless and entirely logical being, and as such, he was held to be a superior decision maker as compared to the emotional humans on his spaceship. But Damasio (1994) has shown that without emotions, it is very difficult for us to make decisions at all! One of Damasio's case studies involved a high-functioning patient (let's call him M) who had developed a brain tumor that pressed on his frontal lobes. The tumor was successfully removed but not without damaging some remaining tissue. After surgery, M appeared to have retained full brain function. But over time, M lost one job after another, got divorced, married and divorced again, and made consistently bad business, personal, and financial decisions. He had, in fact, lost his emotional makeup—and without it, he was unable to mediate and control his rational faculties. For example, M could spend an entire day at work focused on a single detail in a report, unable to evaluate it and move on.

There are many emotions one could investigate. Evans (2001) lists joy, distress, anger, fear, surprise, and disgust as the "basic emotions," with love, guilt, shame, embarrassment, pride, envy, and jealousy rounding out the "higher cognitive emotions." We have looked at anger and disgust, but a case can be made for most of the others vis-à-vis their primacy in human decision making. Rather than continuing down the line with such an exploration; however, let's consider a differently rooted decision driver before we turn to the research on decision making in strategic contexts.

Marketers have mastered a number of ways to confound human decision making, as chronicled in the field of consumer behavior. But until recently, very little has been known about what goes on in our brains when we are manipulated by concepts such as brand appeal or price. McClure et al. (2004) tested a sample of Coke and Pepsi drinkers using fMRI and established that the subjects who had unknowingly preferred Pepsi (half of the participants) had more pronounced brain activity than those preferring Coke—leading to the conclusion that their preferences were stronger when based purely on taste. But for "brand-cued" subjects, who knew which cola they were actually drinking, the Coke drinkers showed very pronounced activity in the ventromedial prefrontal cortex (an area associated with high-level decision-making) which seems to indicate that Coke is, as marketers would say, a "stronger" brand.

Price, it has been shown, also has a significant effect on our neural activity. While previous research (Lee, Frederick, and Ariely, 2006) had established that various manipulations of inputs like brand and ingredients can affect one's reported taste preferences, and even that price manipulation for energy drinks can affect subjects' subsequent performance on puzzles (Shiv, Carmon, and Ariely, 2005), Plassmann et al. (2008) used fMRI scanning for subjects drinking wine. Although the subjects were drinking a single wine in the experiment, they were told otherwise. Believing that the different samples had different prices (e.g., $5 vs. $45), subjects, on the whole, preferred more expensive wines to less expensive ones. The higher prices increased activity in the medial orbitofrontal cortex, an area associated with "experienced pleasantness." Further research might identify other correlates of this phenomenon, for example, whether the higher priced items are believed to confer higher status.

EXPERIMENTAL GAME THEORY

As we have seen, there are numerous ways in which our emotions get involved in our decision making. Prospect theory and, more generally, behavioral decision theory have gone a long way toward identifying patterns in the way that our decisions deviate from those prescribed by rational theory. However, behavioral decision theory has been silent on how we behave in strategic situations. To fill this vacuum, however, we now highlight some results in behavioral game theory: how people actually strategize.

There are some key questions we would like to answer: Do people (in laboratory settings) anticipate the actions of others? Do they actually play dominant strategies? Do they recognize and play mixed strategies? Can they coordinate their moves with others? How much do they trust those other players? How do people handle the tricky Prisoner's Dilemma? And, how do people play when the games are repeated over time? Let's look at the literature.

Consider dominant strategies. Beard and Beil (1994) posed the following two-person, nonzero-sum game to their subjects (Figure 2.4).

Although I have presented this game in strategic form, the game the subjects played was actually a sequential one, with Player 1 moving first. Note that if Player 1 picks the first row, she is guaranteed a payoff of 9.75 units. However, notice also that for Player 2, column 2 weakly dominates column 1. So, Player 2 ought to pick column 2 no matter what, and Player 1, understanding this, will play row 2. Thus, we have found the equilibrium solution (technically, the *subgame perfect equilibrium*; see Selten, 1975), which results in the payoff of (10, 5). Note the risk, however, for Player 1: if Player 2 does not play column 2 as he should, then Player 1 will get stuck with the vastly inferior payoff of 3 had she played row 2 (as she should).

The experimental results deviated from the subgame perfect prescription. Sixty-six percent of the Player 1's chose to play it safe,

	Player 2	
Player 1	(9.75, 3)	(9.75, 3)
	(3, 4.75)	(10, 5)

Figure 2.4   A game of dominance.

sticking with row 1 with a payoff of 9.75 regardless of what the Player 2's did. In other words, only a minority of the Player 1's believed that the Player 2's would play their dominant strategy. However, when the Player 1 participants did play row 2, five out of six of the Player 2's selected their dominant strategy.

Beard and Beil ran some other variations on this game. You might expect that if the risk to Player 1 were reduced, more Player 1's would opt for row 2. Indeed, when decreasing the risk to Player 1, 80% of the Player 1's now picked row 2 (up from 33%). Other experimenters have tested whether people employ dominance (Goeree and Holt, 2001; Camerer, 2003), with similar results: people generally recognize and act on dominance, but are not so sure that others will do so.

Mixed strategies are clearly more difficult to recognize and implement than dominant strategies. Camerer (2003) summarizes a large number of studies and reports that the average of the participants' results, loosely speaking, are weakly correlated with the Nash equilibria. But unlike most of the other game-theoretic properties, mixed strategies have been examined in some real-world settings and this research is very worthwhile to report on. One area of investigation has been penalty kicks in professional soccer leagues. With penalty kicks, the kicker typically shoots left or right. The goalkeeper, who has no way of knowing which way the ball is going to come, must simultaneously dive right or left to have a chance to block the kick. These actions, together with the simple result (goal or no goal) provide a good platform to test whether the players actually employ mixed strategies, and further, whether they are playing minimax (i.e., equilibrium) strategies. Chiappori, Levitt, and Groseclose (2002) and Palacios-Huerta (2003) find that professional soccer players do indeed play minimax. Bar-Eli et al. (2007) find that the optimal strategy for the goalies is to stay in the middle (some kicks, in fact go down the middle—a clever strategy given that goalies always dive to one side or the other); interestingly, the goalies are aware of this, but don't want to appear to the fans as though they are doing nothing!

Given that we have been keeping evolutionary forces in the back of our minds, it should not be surprising to find evidence of mixed strategies in endeavors other than sports. The fields of evolutionary biology and evolutionary psychology provide food for thought. Think Pleistocene: what "strategy" should a woman employ in finding a mate and having children? A mate that would stick around after a baby is born increases the probability that the child will survive. Bearing this in mind, a woman would look for qualities in a partner, indicating that he is willing to invest his time domestically and provide for her and the children in the long run. A stereotypical "bad boy," however, with his good looks but reluctance to commit, would not seem such a good choice in this regard. On the other hand, though, if a woman had a child with an attractive male, the offspring are likely to be attractive and therefore more likely, in turn, to reproduce, thus increasing the chance that the woman's genes will carry on to further generations.

So the question is, should a woman mate with a male who might be less physically attractive but willing to provide for her and the children, or should she mate with a more attractive man, who is likely to run off with someone else in order to maximize *his* reproductive potential? Some fascinating research (see Gangestad, Thornhill, and Garver, 2002; Gangestad and Thornhill, 2008; Flowe, Swords, and Rockey, 2012 and their references) shows that women seem to have developed a mixed strategy that depends on their menstrual cycles. This "cycle shift" hypothesis states that during ovulation (i.e., when they are most fertile), women prefer men whose appearance signals superior genes. These traits include symmetry, social dominance, a more masculine appearance, and increased stature (Flowe, Swords, and Rocky, 2012). In general, the less attractive their current partners, the more pronounced are women's preferences (Pillsworth and Haselton, 2006). Additionally, the converse of this behavior—when women are not fertile, these preferences abate—also holds. Incidentally, some preferred characteristics are not visual—during ovulation, women have an increased attraction to males with low-pitched voices (Puts, 2005; Mlodinow, 2012). It turns out that while there is little correlation between a low-pitched voice and visually masculine characteristics such as height and musculature, there is a significant correlation with testosterone level (Bruckert et al., 2006).

To summarize, the evolutionary-based hypothesis is that women pursue a mixed strategy in their production of offspring: to maximize their chances of producing high-quality offspring that will also survive, they need a partner willing to provide for them in the long term, but they may *also* mate with other males who are likely to possess superior genes.

While the proportion of so-called "extra-pair paternity" in modern populations is low, (below 5% overall, as reported by Gangestad and Thornhill, 2008), the evidence strongly suggests that the preferences exhibited in the recent literature are evolutionary holdovers. The proportion of women who actually act on their observed preferences for

	Player 2	
Player 1	(0, 0)	(200, 600)
	(600, 200)	(0, 0)

Figure 2.5   A coordination game.

dominant men (with symmetrical faces, low-pitched voices, etc.) is unknown, but all indications are that the mixed strategy of sticking with a long-term partner while occasionally selecting a partner with superior genes is a hardwired urge.

Mixed strategies are important, especially for competitive situations, but we often need to cooperate with others. One aspect of cooperation is the ability to coordinate one's actions with others. Consider the nonzero-sum game (Figure 2.5).

In this game, there are two pure Nash equilibria, at (200, 600) and (600, 200). There's also a mixed strategy Nash equilibrium, with the strategy pairs (1/4, 3/4) for both players. Notice that if both players randomize according to the correct mix, they will fail to coordinate (i.e., end up at (0, 0)) 5/8, or 62.5%, of the time. In actual play (Cooper et al., 1978; Camerer, 2003), the results were quite close—a failure rate of 59%. One might hope that with some preplay communication, the players would improve on this result. With so-called one-way communication, where one player declares his move and the other player does not, coordination was reached 96% or the time. The followers in this treatment were able to swallow their pride and accept a payoff of 200 (as opposed to the better outcome of 600, or else 0). But with two-way communication, which simply amounted to the players both declaring their preferences, the players failed to coordinate 42% of the time. It would be interesting to extend this research to explain when and why coordination was achieved.

Another issue that is crucial to our getting along in the world is trust, and it turns out that behavioral game theory has something to say about it. Some insight into our fundamental nature is offered by studies involving the Dictator game. In this game, there are two players and a sum of money, say $20, to be divided between them. Player 1 decides how much money to keep for himself, with the rest of the money going to Player 2. Player 2 performs no action in the game. Initial results, found by Kahneman, Knetsch, and Thaler (1986), were encouraging enough. In their setup, Player 1 had just two options: to split the money 50–50 or to keep $18 out of the available $20. About 75% of the participants opted for the 50–50 split. However, a later experiment by Hoffman, McCabe, and Smith (1996) involved people dividing up the sum of money in a private booth. Over 60% of their subjects took all of the money ($10 in this case), and most of the others just left a dollar or two for the other player. Clearly, society needs some mechanism to deter greed when people are left to their own devices.

The Ultimatum game (Güth, Schmittberger, and Schwarze, 1982) calibrates just how much people try to get away with when their choices are contested by others. This game is a Dictator game in which Player 2 can either accept or reject the action of Player 1. If Player 2 rejects Player 1's offer, then both players get nothing. With a total available of $10, for example, where the units are in increments of $0.01, the (subgame perfect) Nash equilibrium of the game is for Player 1 to offer $0.01 to Player 2, and for Player 2 to accept. But virtually no one plays the equilibrium strategy; furthermore, Player 2's who are offered small amounts are insulted and prefer to decline, thereby punishing greedy Player 1's. Players 1 seem to anticipate this behavior. Camerer (2003) summarizes the copious literature: the average and median amounts offered are in the neighborhood of 40% of the pot; these typical offers of 35–50% are accepted the great majority of the time, while offers in the 20% range are rejected about half the time. Clearly, Player 1's in the Ultimatum game are wary (and they should be!) that the Player 2's might reject a low offer.

Another type of trust experiment looks at sums of money being exchanged back and forth between players. In the so-called Centipede game, there are two players and an initial pot of money. Player 1 can initially "take" a certain percentage of the pot (leaving the rest for Player 2) or "pass." If Player 1 passes, the pot doubles and Player 2 is left with the same set of options. If Player 2 plays pass, the pot once again doubles and it is again Player 1's turn. Once a player plays take, the pot is divided accordingly and the game ends. Using backward induction, it is easy to show that the subgame perfect equilibrium in this game is for Player 1 to play take on the first move. McKelvey and Palfrey (1992) studied a four-round Centipede game with an initial pot of $0.50 (see Fig. 2.6).

In their study, fewer than 10% of the players played take on the first move, and, surprisingly, about 20% of the Player 2's who reached the

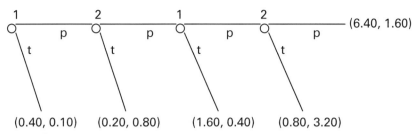

Figure 2.6   A four-round Centipede game. Payoffs in dollars. p, pass; t, take.

fourth and last round played pass—forgoing a payoff of $3.20 to instead accept $1.60 (giving $6.40 to the other player). Clearly, despite the risk that the other player might play take on the next move, the great majority of the subjects in this study understood that it is better to grow the pot as opposed to playing the rational strategy. And the fact that so many players played pass does not just reflect their economic savvy, but is also a testament to their trust in the other player being willing to grow the pot on the next move.

The Prisoner's Dilemma also involves a measure of trust. The first PD experiment occurred on the very day that the game was conceived, in 1950. As we know, the unique pure strategy equilibrium occurs when both players defect. But since many of the social and commercial situations that the PD models are in fact repeated over time, it makes sense to iterate the game in an experimental setting. The rational strategy in a finitely repeated PD would be one of serial defection. But, according to Poundstone (1992), when the game was repeated 100 times between 2 well-known scholars back on that fateful day in 1950, both of them cooperated most of the time (68 and 78 times out of the 100 plays). The humans playing the game recognized that they could string along a series of mutually cooperative plays, thereby yielding a significantly higher long-term average payoff than they would obtain by serially defecting.

Sally (1995) provides a meta-analysis of 30 PD experiments. The mean cooperation rate overall was 47.4%, with a high standard deviation across the experimental treatments. While this overall rate is not as high as in the initial experiment, it is clear that human subjects often reject the defection strategy. Kreps et al. (1982) developed a game-theoretic model to justify such behavior. In this model, each player believes that there is some probability $p > 0$ that the *other* player is irrational, that is, would cooperate in any one play of the game. Given this assumption, the theoretical result provides an equilibrium in which both players will serially cooperate (although not in the final round). This result is consistent with the so-called Folk theorem (see Fudenberg and Tirole, 1991), which states that in a repeated game, if future payoffs are high enough, then any series of plays that yields an average payoff at least as good as the minimax outcome can be maintained by the players and is therefore a subgame perfect equilibrium. For the repeated PD, this means that mutual cooperation can end up as a Nash equilibrium if the players perceive future interactions as either likely enough or valuable enough.

Andreoni and Miller (1993) had subjects play the PD against either partners or strangers. The idea was to see whether, indeed, players would play toward developing a reputation with opponents that they met repeatedly, as opposed to strangers for whom the notion of reputation is moot. When playing strangers, the participants cooperated an average of about 20% of the time; but when playing against the steady partners, the cooperation rate rose, on the average, to just over 40%. Another interesting finding is that the players could fairly neatly be divided into three types: cooperators, defectors, and mixers. Defectors almost always defected; cooperators generally cooperated most of the way through the plays; mixers cooperated with some probability p, which on average was about 20%.

Frank, Gilovich, and Regan (1993) instead studied one-shot PDs, with an eye toward discovering whether they could increase cooperation through preplay communication as opposed to reputation building through repeated play. They divided their subjects into groups of three players each. Each subject played once each against the others in his group. But prior to actual play, each group spent a half hour alone together. The idea was that getting to know one another would not only foster cooperation but also enable the players to gauge how likely they believed the others would cooperate. Out of a total of 198 plays overall, the subjects cooperated (C) 146 times (74%). Also, the subjects correctly predicted 130 of the 146 Cs and 31 of the 52 defections, an accuracy rate of 76%.

Moving forward, what else can we learn from games such as the Ultimatum game and the PD? Rilling et al. (2002) performed fMRI scanning on women who played the PD against other women. Their results showed that mutual cooperation was associated with increased neural activity in a variety of areas (the nucleus accumbens, the caudate nucleus, the ventromedial frontal cortex, and the rostral anterior cingulate cortex) that are active in reward processing. Beyond cooperation in the PD, it appears that these areas of the brain reinforce (or are reinforced by) general behaviors of reciprocal altruism. Scientists are beginning to study the neural correlates of altruism, and while thus far there are no definitive results, Tankersley, Stowe, and Huettel (2007) have identified neural correlates of perceived "agency" by others on their subjects' behalf.

With the Ultimatum game, Sanfey et al. (2003) used fMRI scanning to focus on the subjects as they reacted to "fair" or "unfair" proposals. Unfair proposals (i.e., where the Player 2's were offered from 10 to 30% of the $10 pot) were met with increased activity in the anterior insula and the dorsolateral prefrontal cortex, areas associated with emotion and reflection, respectively. It would appear that the anterior insula is generating disgust, while the dorsolateral prefrontal cortex is trying to coax the subject into taking the money. The higher the activity in the anterior insula, the more likely the subjects were to reject the unfair offers. A study by Paulus et al. (2003) found that when subjects considered risky alternatives in general, the degree of risk was positively correlated with the degree of activity in the insular cortex. Interestingly, the level of activity in the insular cortex was in turn related to the subject's propensity for avoiding harm, and for neuroticism, as measured on personality scales.

One important feature of the Sanfey et al. (2003) study is that the conflict between our emotional and rational "selves" was localized in the anterior insula and dorsolateral prefrontal cortex, respectively, as seen in the fMRI scans. It is instructive to consider an additional such skirmish before we move on. Westen et al. (2006) recruited strongly committed male Democrats and Republicans during the 2004 US presidential campaign to answer a series of questions while they were being scanned with fMRI. First, they were presented a statement by one of the candidates, followed by a statement showing that the candidate's statement was inconsistent. Then, the subjects were presented with questions where they had to rate whether the candidate's statements were inconsistent, and to what degree. This was followed by a statement that "explained away" the inconsistency. The final two questions asked the subjects to consider whether the candidate's statements were actually less inconsistent than they initially thought, and by how much.

Not surprisingly, the Democrats gave the Democratic candidate (John Kerry) a break but not the Republican candidate (George W. Bush), while the Republicans gave Bush a break but not Kerry. But the point of the study was to locate the neural correlates of such "motivated" reasoning, where we adopt justifications that minimize cognitive dissonance. Such reasoning is related to the confirmation bias that we discussed earlier. The main finding was that motivated reasoning is different from cold, logical reasoning (when our emotions are not engaged). More specifically, Westen et al. discovered that when the subjects were presented with contradictory (and therefore, emotionally threatening) statements from their favored candidate, they arrived at an alternative conclusion (i.e., cutting their candidate a break), and in developing this judgment, the "cold," rational parts of their brains were not active, but the emotional lobes, such as the insula, were.

BEHAVIOR MODIFICATION AND CONCLUSIONS

We've covered a lot of ground thus far, and now we should stop and ask, where is this taking us? In particular, is there some way for us, as individuals and as a society, to harness what we have learned in a constructive manner?

Thaler and Sunstein in their book *Nudge* (2008) have tried to provide a roadmap for how we can redirect ourselves away from some of the more unfortunate paths that our biased thought processes lead to. Their idea is to promote a philosophy, called *libertarian paternalism*, which we can apply to a number of both personal and public sector problems. Given a certain situation in which decision-makers would choose from a number of alternatives, how can one design a "choice architecture" that points, or "nudges," them down a path that makes them better off (from their points of view) while still leaving them free to choose?

Let's look at a couple of their examples. We already know from prospect theory that alternative "framing" of certain decisions—for example, when comparing outcomes to different reference points—can change the way we view them. The process to become an organ donor is a good example of this principle in action. In most states in the United States, one has to take a concrete action, that is, to explicitly provide consent, in order to donate one's organs. But while the great majority of us believe that being an organ donor is a good thing, far fewer of us have checked the box on our driver's licenses. In other words, when the default policy is that we are not donors, far fewer of us participate as opposed to a system where the consent is presumed—where we would have to take action to opt out.

Another example is the way that social security was privatized in Sweden. More than a decade ago, the Swedish government overhauled their national pension fund. There were a number of design choices they faced in turning over individual pension fund control to their citizens: at one extreme, participants would be forced into a single default fund; at the other extreme, participants would be forced to select their allocations or else get dropped from the fund altogether; and in the middle, a default fund would be available, but the government would also promote, to some varying degree, free choice of asset allocation.

The Swedish government chose a policy whereby a default fund was made available to participants but selecting it was discouraged. At first sight, encouraging participants to seek out their own asset allocations seems to be a reasonable approach. But (with the help of hindsight), Thaler and Sunstein report that on the whole, people who chose their own investments did much worse than those enrolled solely in the default plan. While the moral of this story is that it can be dangerous to let Humans have such free choice, it remains unclear in general as to when people should be reined in (protected from the consequences of their biases) or not.

There are different ways in which people can be supported in their quest to improve their lives. Recently, a number of behavior-modification programs have appeared, which are designed to help people lose weight, curb impulsive purchasing behavior, and break bad habits in general (Freedman, 2012). These programs, typically run as smartphone apps, are gaining in both popularity and efficacy. While one can quibble about whether behavior modification itself seems to treat us as automata as opposed to willful agents, some of the success of these programs indicates that we can, as Freedman says, "purposefully alter our environment to shape our behavior in ways we choose."

You already knew that people make some decisions without even knowing—like the folks in movie theaters who were shown subliminal messages to drink Coke, and, unaware of their subconscious desires, went straight to the concession stand. But what is remarkable is the sheer scope and variety of our emotional decision-making apparatus, and the difficulties our deliberate sides have in overcoming the gut feelings. It is important to observe, as Lynch (2012) does, that over time, as individuals and as a society we have done just this, in the decades-long shift in our attitudes on issues like race, environmentalism, and climate change.

Indeed, one is tempted to sound a "call to arms" to pursue a research program in which we aim to identify what categories of decisions are best left to our automatic systems, and which ones are appropriate for our reflective systems. The research we have seen on the IAT (Greenwald, McGhee, and Schwartz, 1998) and also on the affect heuristic (Finucane et al., 2000) might encourage us to believe that given enough time, our rational "override buttons" can be widely employed to overcome our biased, gut reactions. However, we have also seen in the Westen et al. study (2006) that sometimes the override button is operated by our emotional side, and having that extra bit of time to process our decision leads us down the wrong path. Such "moral dumbfounding" (Haidt and Hersh, 2001) certainly complicates the picture as we move forward; and as if that isn't enough, let's not forget that the evolutionary honing of our emotions further calls into question whether our reflective systems should always be trusted! Certainly, the question of determining when and why to use our rational "override" buttons will be a topic of central interest in the social sciences.

Ultimately, to navigate the world like Odysseus, we will need to learn more about ourselves than we currently do. Who among us can be nudged? Under what circumstances will nudging be successful? In how many ways can we positively modify our behavior? On the individual level, perhaps we will soon know enough about our genetic makeup to be able to customize programs that will enhance numerous aspects of our lives. And on the societal stage, maybe someday soon, we will understand how to better integrate our hearts and minds, thereby helping ourselves and our future generations to make wiser choices than the ones we make today.

REFERENCES

Ainslie, G.W., "Impulse control in pigeons," *Journal of the Experimental Analysis of Behavior* 21, 3 (1974) 485–489.

Ainslie, G.W. and R.J. Herrnstein, "Preference reversal and delayed reinforcement," *Animal Learning Behavior* 9, 4 (1981) 476–482.

Allais, M., "Le comportement de l'homme rationnel devant le risque: Critique des postulats et axiomes de l'ecole Americaine," *Econometrica* 21, 4 (1953) 503–546.

Andreoni, J. and J. Miller, "Rational cooperation in the finitely repeated prisoner's dilemma: Experimental evidence," *The Economic Journal* 103 (1993) 570–585.

Ariely, D., Loewenstein, G., and D. Prelec, "Coherent arbitrariness: Stable demand curves without stable preferences," *Quarterly Journal of Economics* 118, 1 (2003) 73–106.

Bar-Eli, M., Azar, O.H., Ritov, I., Keidar-Levin, Y., and G. Schein, "Action bias among elite soccer goalkeepers: The case of penalty kicks," *Journal of Economic Psychology* 28, 5 (2007) 606–621.

Beard, T.R. and R.O. Beil, "Do people rely on the self-interested maximization of others? An experimental test," *Management Science* 40, 2 (1994) 252–262.

Bernoulli, D., (1738) "Exposition of a new theory of the measurement of risk," *Econometrica* 22 (1954) 23–36 (English translations).

Binmore, K., *Playing for Real: A Text on Game Theory*. Oxford University Press, New York and Oxford (2007).

Bruckert, L., Liénard, J.-S., Lacroix, A., Kreutzer, M., and G. Leboucher, "Women use voice parameters to assess men's characeristics," *Proceedings of the Royal Society B* 273 (2006) 83–89.

Camerer, C.F., *Behavioral Game Theory: Experiments in Strategic Interaction*. Russell Sage Foundation, New York (2003).

Chiappori, P.-A., Levitt, S., and T. Groseclose, "Testing mixed-strategy equilibria when players are heterogeneous: The case of penalty kicks in soccer," *American Economic Review* 92, 4 (2002) 1138–1151.

Cooper, R.W., DeJong, D.V., Forsythe, R., and T.W. Ross, "Selection criteria in coordination games: Some experimental results," *American Economic Review* 80, 1 (1978) 218–233.

Damasio, A., *Descartes Error: Emotion, Reason, and the Human Brain*. Penguin, New York (1994).

Dutta, P.K., *Strategies and Games: Theory and Practice*. MIT Press, Cambridge, MA (1999).

Evans, D., *Emotion*. Oxford University Press, Oxford (2001).

Finucane, M.L., Alhakami, A., Slovic, P., and S.M. Johnson, "The affect heuristic in judgment of risks and benefits," *Journal of Behavioral Decision Making* 13, 1 (2000) 1–17.

Fitzgerald, D.A., Posse, S., Moore, G.J., Tancer, M.E., Nathan, P.J., and K.L. Khan, "Neural correlates of internally-generated disgust via autobiographical recall: A functional magnetic resonance imaging investigation," *Neuroscience Letters* 370 (2004) 91–96.

Flowe, H.D., Swords, E., and J.C. Rockey, "Women's behavioural engagement with a masculine male: Evidence for the cycle shift hypothesis," *Evolution and Human Behavior* 33, 4 (2012) 285–290.

Frank, R.H., *Passions within Reason: The Strategic Role of the Emotions*. W.W. Norton & Company, New York (1988).

Frank, R., Gilovich, T., and D. Regan, "The evolution of one-shot cooperation: An experiment," *Ethology and Sociobiology* 14, 4 (1993) 247–256.

Freedman, D.H., "The perfected self," *The Atlantic* (June 2012), http://www.theatlantic.com/magazine/archive/2012/06/the-perfected-self/308970/ (accessed August 22, 2014).

Fudenberg, D. and J. Tirole, *Game Theory*. MIT Press, Cambridge, MA (1991).

Gangestad, S.W. and R. Thornhill, "Human oestrus," *Proceedings of the Royal Society B* 275 (2008) 991–1000.

Gangestad, S.W., Thornhill, R., and C.E. Garver, "Changes in women's sexual interests and their partners' mate-retention tactics across the menstrual cycle," *Proceedings of the Royal Society B* 269 (2002) 975–982.

Goeree, J.K. and C.A. Holt, "Ten little treasures of game theory and ten intuitive contradictions," *American Economic Review* 91 (2001) 1402–1422.

Greene, J.D. and J. Haidt, "How (and where) does moral judgment work?" *Trends in Cognitive Sciences* 6, 12 (2002) 517–523.

Greene, J.D., Sommerville, R.B., Nystrom, L.E., Darley, J.M., and J.D. Cohen, "An fMRI investigation of emotional engagement in moral judgment," *Science* 293 (2001) 2105–2108.

Greenwald, A.G., McGhee, D.E., and J.L.K. Schwartz, "Measuring individual differences in implicit cognition: The implicit association test," *Journal of Personality and Social Psychology* 74, 6 (1998) 1464–1480.

Güth, W., Schmittberger, R., and B. Schwarze, "An experimental analysis of ultimatum bargaining," *Journal of Economic Behavior and Organization* 3, 4 (1982) 367–388.

Haidt, J., "The moral emotions," in Davidson, R.J., Scherer, K.R., and H.H. Goldsmith (Eds.), *Handbook of Affective Sciences*. Oxford University Press, Oxford (2003).

Haidt, J. and M.A. Hersh, "Sexual morality: The cultures and emotions of conservatives and liberals," *Journal of Applied Social Psychology* 31, 1 (2001) 191–221.

Hoffman, E., McCabe, K.A., and V.L. Smith, "On expectations and the monetary stakes in ultimatum games," *International Journal of Game Theory* 25, 3 (1996) 289–301.

James, W., "What is an emotion," *Mind* 9, 34 (1884) 188–205.

Johnson, M.W., Bickel, W.K., Baker, F., Moore, B.A., Badger, G.J., and A.J. Budney, "Delay discounting in current and former marijuana-dependent individuals," *Experimental and Clinical Psychopharmacology* 18, 1 (2010) 99–107.

Kable, J.W. and P.W. Glimcher, "The neural correlates of subjective value during intertemporal choice," *Nature Neuroscience* 10 (2007) 1625–1633.

Kahneman, D., *Thinking, Fast and Slow*. Farrar, Straus and Giroux, New York (2011).

Kahneman, D. and A. Tversky, "Prospect theory: An analysis of decision making under risk," *Econometrica* 47, 2 (1979) 263–292.

Kahneman, D., Knetsch, J.L., and R. Thaler, "Fairness and the assumptions of economics," *The Journal of Business* 59, 4 (1986) S285–S300.

Kreps, D.M., Milgrom, P., Roberts, J., and R. Wilson, "Rational cooperation in the finitely repeated prisoner's dilemma," *Journal of Economic Theory* 27, 2 (1982) 245–252.

Lee, L., Frederick, S., and D. Ariely, "Try it, you'll like it: The influence of expectation, consumption, and revelation on preferences for beer," *Psychological Science* 17, 12 (2006) 1054–1058.

Libet, B., "The experimental evidence for subjective referral of a sensory experience backwards in time: Reply to P.S. Churchland," *Philosophy of Science* 48, 2(1981) 182–197.

Libet, B., Wright, E.W. Jr., and C.A. Gleason, "Preparation- or intention-to-act, in relation to pre-event potentials recorded at the vertex," *Electroencephalography and Clinical Neurophysiology* 56, 4 (1983) 367–372.

Loewenstein, G., Read, D., and R.F. Baumeister (Eds.), *Time and Decision*. Russell Sage Foundation, New York (2003).

Lynch, M. P., *In Praise of Reason*. MIT Press, Cambridge, MA (2012).

McClure, S., Li, J., Tomlin, D., Cypert, K., Montague, L.M., and P. Montague, "Neural correlates of behavioral preference for culturally unfamiliar drinks," *Neuron* 44, 2 (2004) 379–387.

McKelvey, R.D. and T.R. Palfrey, "An experimental study of the centipede game," *Econometrica* 60, 4 (1992) 803–836.

Mlodinow, L., *Subliminal: How Your Unconscious Mind Rules Your Behavior*. Pantheon Books, New York (2012).

Moll, J., de Oliveira-Souza, R., Eslinger, P.J., Bramati, I.E., Mourão-Miranda, J., Andreiuolo, P.A., and L. Pessoa, "The neural correlates of moral sensitivity: A functional magnetic resonance imaging investigation of basic and moral emotions," *The Journal of Neuroscience* 22, 7 (2002) 2730–2736.

Mudambi, S.M., "The games retailers play," *Journal of Marketing Management* 12 (1996) 695–706.

Myerson, R.B., *Game Theory*. Harvard University Press, Cambridge, MA (1991).

Nash, J., "Equilibrium points in n-person games," *Proceedings of the National Academy of Sciences* 36 (1950) 48–49.

Nosek, B.A., Greenwald, A.G., and M.R. Banaji, "The implicit association test at age 7: A methodological and conceptual review," in J.A. Bargh (Ed.), *Automatic Processes in Thinking and Behavior*. Psychology Press, London (2007).

Palacios-Huerta, I., "Professionals play minimax," *Review of Economic Studies* 70, 2 (2003) 395–415.

Parkhe, A., Rosenthal, E.C., and R. Chandran, "Prisoner's dilemma payoff structure in interfirm strategic alliances: An empirical test," *Omega International Journal of Management Science* 21, 5 (1993) 531–539.

Paulus, M.P., Rogalsky, C., Simmons, A., Feinstein, J.S., and M.B. Stein, "Increased activation in the right insula during risk-taking decision making is related to harm avoidance and neuroticism," *NeuroImage* 19, 4 (2003) 1439–1448.

Pillsworth, E.G. and M.G. Haselton, "Male sexual attractiveness predicts differential ovulatory shifts in female extra-pair attraction and male mate retention," *Evolution and Human Behavior* 27, 4 (2006) 247–258.

Plassmann, H., O'Doherty, J., Shiv, B., and A. Rangel, "Marketing actions can modulate neural representations of experienced pleasantness," *Proceedings of the National Academy of Sciences* 105, 3 (2008) 1050–1054.

Poundstone, W., *Prisoner's Dilemma*. Doubleday, New York (1992).

Puts, D.A., "Mating context and menstrual phase affect women's preferences for male voice pitch," *Evolution and Human Behavior* 26 (2005) 388–397.

Quiggin, J., "A theory of anticipated utility," *Journal of Economic Behavior and Organization* 3, 4 (1982) 323–343.

Rapoport, A. and A.M. Chammah, *Prisoner's Dilemma: A Study in Conflict and Cooperation*. University of Michigan Press, Ann Arbor (1965).

Rilling, J.K., Gutman, D.A., Zeh, T.R., Pagnoni, G., Berns, G.S., and C.D. Kilts, "A neural basis for social cooperation," *Neuron* 35, 2 (2002) 395–405.

Rozin, P., Haidt, J., and C.R. McCauley, "Disgust," in Lewis, M., Haviland-Jones, J.M., and L.F. Barrett (Eds.), *Handbook of Emotions* (3rd ed.). Guildford Press, New York (2008).

Sally, D., "Conversation and cooperation in social dilemmas: A meta-analysis of experiments from 1958 to 1992," *Rationality and Society* 7, 1 (1995) 58–92.

Sanfey, A.G., Rilling, J.K, Aronson, J.A., Nystrom, L.E., and J.D. Cohen, "The neural basis of economic decision-making in the ultimatum game," *Science* 300 (2003) 1755–1758.

Sartre, J.-P., (1943) *Being and Nothingness: An Essay in Phenomenological Ontology*, trans. by Hazel Barnes. Citadel Press, New York (1965).

Savage, L.J., *The Foundation of Statistics*. John Wiley & Sons, Inc., New York (1954).

Schmeidler, D., "Subjective probability and expected utility without additivity," *Econometrica* 57, 3 (1989) 571–587.

Selten, R., "Reexamination of the perfectness concept for equilibrium points in extensive games," *International Journal of Game Theory* 4, 1 (1975) 2–55.

Shih, M., Pittinsky, T.L., and N. Ambady, "Stereotype susceptibility: Identity salience and shifts in quantitative performance," *Psychological Science* 10, 1 (1999) 80–83.

Shiv, B., Carmon, Z., and D. Ariely, "Placebo effects of marketing actions: Consumers may get what they pay for," *Journal of Marketing Research* 42, 4 (2005) 383–393.

Spinoza, B. (1677), *The Ethics and selected letters*, trans. by Samuel Shirley. Hackett Publishing Co., Indianapolis (1982).

Stanovich, K.E. and R.F. West, "Individual differences in reasoning: Implications for the rationality debate?" *Behavioral and Brain Sciences* 23 (2000) 645–726.

Tankersley, D. Stowe, C.J., and S.A. Huettel, "Altruism is associated with an increased neural response to agency," *Nature Neuroscience* 10 (2007) 150–151.

Thaler, R., "Some empirical evidence on dynamic inconsistency," *Economics Letters* 8 (1981) 201–207.

Thaler, R.H. and C.R. Sunstein, *Nudge*. Penguin Books, New York (2008).

Tversky, A. and D. Kahneman, "Extensional versus intuitive reasoning: The conjunction fallacy in probability judgment," *Psychological Review* 90 (1983) 293–315.

Tversky, A. and D. Kahneman, "Advances in prospect theory: Cumulative representation of uncertainty," *Journal of Risk and Uncertainty* 5, 4 (1992) 297–323.

von Neumann, J., "Zur theorie der gesellschaftsspiele," *Mathematische Annalen* 100 (1928) 295–320.

von Neumann, J. and O. Morgenstern, *Theory of Games and Economic Behavior*. Princeton University Press, Princeton, NJ (1944).

Wakker, P. and D. Deneffe, "Eliciting von Neumann-Morgenstern utilities when probabilities are distorted or unknown," *Management Science* 42, 8 (1996) 1131–1150.

Wakker, P. and A. Tversky, "An axiomatization of cumulative prospect theory," *Journal of Risk and Uncertainty* 7, 2 (1993) 147–175.

Weber, B.J. and S.A. Huettel, "The neural substrates of probabilistic and intertemporal decision making," *Brain Research* 1234 (2008) 104–115.

Wegner, D., *The Illusion of Conscious Will*. MIT Press, Cambridge, MA (2002).

Wertenbroch, K., "Self rationing: Self control in consumer choice," in Loewenstein, G., Read, D., and R.F. Baumeister (Eds.), *Time and Decision: Economic and Psychological Perspectives on Intertemporal Choice*. Russell Sage Foundation, New York (2003).

Westen, D., Blagov, P.S., Harenski, K., Kilts, C., and S. Hamann, "Neural bases of motivated reasoning: An fMRI study of emotional constraints on partisan political judgment in the 2004 U.S. presidential election," *Journal of Cognitive Neuroscience* 18, 11 (2006) 1947–1958.

Yaari, M.E., "The dual theory of choice under risk," *Econometrica* 55, 1 (1987) 95–115.

Zajonc, R.B., "Feeling and thinking: Preferences need no inferences," *American Psychologist* 35, 2 (1980) 151–175.

CHAPTER 3

Simulation Optimization: Improving Decisions under Uncertainty

Marco Better[1], Fred Glover[1], and Gary Kochenberger[2]
[1] *OptTek Systems, Inc., Boulder, CO, USA*
[2] *Business Analytics, School of Business, University of Colorado-Denver, Denver, CO, USA*

INTRODUCTION

Analytics has been defined as "the scientific process of transforming data into insight for making better decisions." More and more organizations are using analytics to make better decisions and reduce risks. Analytics includes well-established methods such as mathematical optimization, simulation, probability theory, and statistics, as well as newer techniques that take elements from traditional methods and modify and/or combine them into robust frameworks in order to develop more powerful solution methods for many settings where traditional methods fall short. A prime example of the latter is the simulation–optimization framework. As its name implies, this method combines simulation and optimization in order to tackle complex situations where risk and uncertainty do not behave according to certain simplifying assumptions.

Breakthroughs in Decision Science and Risk Analysis, First Edition.
Edited by Louis Anthony Cox, Jr.

Taken separately, each method is critical, but limited in scope. On the one hand, optimization by itself provides an excellent method to select the best element in terms of some system performance criteria, from some set of available alternatives, in the absence of uncertainty. On the other hand, simulation is a tool that allows us to build a representation of a complex system in order to better understand the uncertainty in the system's performance.

By putting these two methods together, we can develop a powerful framework that takes advantage of each method's strengths, so that we have at our disposal a technique that allows us to select the best element from a set of alternatives and simultaneously take account of the uncertainty in the system.

In this chapter, we will begin with an example to illustrate each technique separately, and highlight the benefits of simulation optimization approaches in the presence of uncertainty. We will then explore the use of simulation optimization in real-world applications in risk-management. Finally, we will summarize our discussion in a set of concluding remarks.

AN ILLUSTRATIVE EXAMPLE

"Portfolio investment theory" concerns finding the portfolio of investments that maximizes expected returns while minimizing risk. Since investors are risk-averse, they prefer portfolios with high expected returns and low risk (Sharpe, 1964).

In 1952, Nobel laureate Dr Harry Markowitz laid down the foundation for modern investment theory. Markowitz focused his attention on *mean-variance efficient* portfolios (Markowitz, 1952). A portfolio is *mean-variance efficient* if it has the highest expected (mean) return for a given variance, or, similarly, if it has the smallest variance for a given expected return. Markowitz developed this theory for portfolios of securities, such as stocks and bonds, for which returns are usually normally distributed. If portfolio returns are normally distributed, then its risk can be measured by the variance of its returns. If this is the case, then the *optimal* set of portfolios can be found by traditional optimization methods, as we show in the following section.

OPTIMIZATION OF SECURITIES PORTFOLIOS

What constitutes the best portfolio of securities? In 1952, Markowitz attempted to answer this question with his famous paper, "Portfolio Selection." In this paper, Markowitz formulated the portfolio selection problem as an optimization problem.

For simplicity, let's first consider a market that contains only three assets (i.e., stocks), A_1, A_2, and A_3. Let's assume that we have a limited budget to invest in these assets, and that we want to invest our budget in its entirety among these assets. Therefore, we will denote x_1, x_2, and x_3 as the proportion of our budget that we will invest in A_1, A_2, and A_3, respectively. Since we will invest our entire budget, it follows that:

$$x_1 + x_2 + x_3 = 1$$

Now, let μ_1, μ_2, and μ_3 denote the expected value (i.e., the mean) of the return of A_1, A_2, and A_3, respectively; and let $\sigma_{(1)}^2$, $\sigma_{(2)}^2$, and $\sigma_{(3)}^2$ denote the variance of the probability distribution of the returns of A_1, A_2, and A_3, respectively. Thus, we will use $\sigma_{(1,2)}^2$ to represent the covariance between the probability distributions of the returns of A_1, A_2, and so forth.

The mean returns, return variances, and covariances can be estimated from historic stock price data.

The mean return of the portfolio, μ_p, will be equal to the weighted average expected return, where the weights correspond to the proportion of the budget invested in each asset; thus,

$$\mu_p = \mu_1 x_1 + \mu_2 x_2 + \mu_3 x_3$$

The variance of the probability distribution of the portfolio returns, $\sigma_{(p)}^2$, is calculated as follows:

$$\sigma_{(p)}^2 = \sigma_{(1,2)}^2 x_1 x_2 + \sigma_{(1,3)}^2 x_1 x_3 + \sigma_{(2,3)}^2 x_2 x_3$$

Therefore, we can formulate the problem of finding the best portfolio as follows:

$$\text{Maximize} \quad \left(\mu_p - \sigma_{(p)}^2 \right) \tag{3.1}$$

$$\text{Subject to:} \quad x_1 + x_2 + x_3 = 1 \tag{3.2}$$

$$x_1, x_2, x_3 \geq 0 \tag{3.3}$$

Equation 3.1 is known as the objective function, because it represents our main goal, or objective. In this case, our objective is to find the portfolio with the maximum expected return and minimum risk.

Equation 3.2 makes sure the entire budget is invested among the assets; and Equation 3.3 enforces a positivity constraint on our investments (no short positions are allowed).

We could have used the alternate objective function:

$$\text{Minimize} \quad \left(\sigma^2_{(p)} - \mu_p \right) \tag{3.1'}$$

In this case, our objective is to find the portfolio with minimum risk and maximum return. It is apparent that Equations 3.1 and are equivalent.

These formulations can be solved using a specific type of mathematical programming technique called "quadratic programming." By using quadratic programming, we can find the optimal solution(s) to the above portfolio selection problem. Unfortunately, there are two complicating factors here:

1. There may be many—in certain cases, even an infinite number of—optimal solutions to the formulation mentioned earlier. In fact, Markowitz calls the set of solutions the "efficient set," and they all lie along a curve called the "efficient frontier." It is up to the investor to "pick" the solution on the frontier that provides maximum return for the level of risk she is willing to accept.

2. The formulation is valid only if one very strict assumption holds: the returns of each of the assets, and, hence, the portfolio returns must follow a normal probability distribution. If this is not the case, then the Markowitz model breaks down. Under these strict normality conditions the variance is symmetric, so that the probability that the actual portfolio return will be above its estimated mean is the same as the probability that it will be below. However, using the variance as the only risk measure does not protect us from the probability that the portfolio return will be very small, or even negative; in other words, we have no idea of the extent to which our investment is "at risk."

In most practical situations, especially when there is a large number of underlying assets to be considered, these complicating factors make any solution found by this method unreliable. On the one hand, it is difficult to quantify the exact level of risk a particular investor is willing to accept.

On the other hand, portfolio returns often do not follow a normal distribution, or the distribution is severely skewed when investment costs and capital gains tax implications come into play. Therefore, although we have a very elegant model like Markowitz's, we cannot use it.

We need a method that can provide us with more complete information about the quality of the portfolio in terms of risk and return.

SIMULATION

Monte Carlo simulation is a method used by financial companies to simulate and understand risks related to various investments (Hertz, 1979). The main advantage of this method is that the normality assumption is no longer a requirement; in fact, the power of the method is that we can use statistical techniques to analyze an asset's historical data, and forecast its behavior into the future by simulating the probable outcomes. This provides freedom from strict assumptions about the probability distribution of the assets.

To illustrate this better, we have taken a sample of month-end closing stock prices for six very well-known high-tech corporations: Sun Microsystems, Oracle, Microsoft, Intel Corporation (INTEL), Cisco Systems, and Yahoo, for the period between March 31, 1999, and April 30, 2001.

Let's take for instance, the end-of-month closing stock price for INTEL. During the (roughly) 2-year period, INTEL's end-of-month stock price averaged $75.50, with an average monthly change of −2.74 percentage points, and an overall drop of 74% from $118.88 to $30.91. During that period, the month-end stock price increased 11 times and decreased 14 times. The biggest month-to-month change for INTEL during that period was a 48.5% drop, from a $118.88 closing price on March 31, 1999 to April 30, 1999; conversely, the biggest increase of 23.1%, from a closing price of $30.03 to $37.00 was registered between December 30, 2000, and January 30, 2001.

Using this information and some additional statistics, we can find a probability distribution that these data fit quite well. Several commercial software products exist that can perform "goodness-of-fit" testing for data such as these. For our example, we used Crystal Ball's "Batch Fit" feature, which automatically finds the best fit among a set of more than a dozen well-known probability distributions (see http://www.oracle.

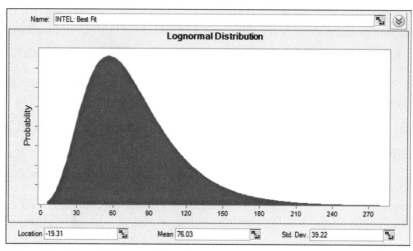

Figure 3.1 Crystal Ball screen shot of lognormal probability distribution.

com/us/products/applications/crystalball for more information about the Batch Fit and other features).

According to Crystal Ball, the historic end-of-month stock price best fits a lognormal probability distribution with mean = $76.03 and standard deviation = $39.22. If we plot this distribution, we obtain a graph like the one shown in Figure. 3.1.

From Figure 3.1, we can deduce that there is an 80% likelihood that month-end stock prices for INTEL will fall between roughly $33.00 and $127.00, with only a 20% chance of being higher or lower.

Similarly, we can find the best-fitting distributions for each of the stocks that we are interested in including in our portfolio. These results are summarized in Table 3.1.

Based on information contained in the third column and the last column of Table 3.1, which contain the standard deviation and the average monthly return for each stock, if we were to use the pure optimization approach discussed previously, the optimal portfolio would result in an average monthly return of 2.59% over the next 2 years, with a standard deviation of returns of 0.19%, by taking the following portfolio positions:

- Invest about 25.7% of our budget in ORACLE stock;
- Invest about 17.7% of the budget in MICROSOFT stock;
- Invest about 47.5% of the budget in INTEL stock; and,
- Invest the remaining 9.1% of the budget in CISCO stock.

Table 3.1 Summary statistics for selected stock prices

Stock	Mean stock price	Standard deviation	P(10)	P(90)	Selected distribution	Average monthly return (%)
Sun Microsystems	$77.96	$33.84	$30.99	$117.57	Min extreme	-3.69
Oracle	49.93	26.85	18.44	89.90	Beta	2.93
Microsoft	78.33	17.01	56.59	100.62	Beta	0.39
Intel Corporation	76.03	35.79	33.82	127.04	Lognormal	-2.74
Cisco Systems	68.80	30.52	30.81	108.86	Lognormal	-5.18
Yahoo	130.30	93.48	21.68	238.91	Logistic	-3.74

However, if we simulate this portfolio using Monte Carlo simulation and the best-fit distribution information given earlier, we obtain the following results:

- Average monthly portfolio return = 0.99%
- Standard deviation of returns = 0.93%

We can see that the simulation of the portfolio, which takes into account the uncertainty in the stock prices (i.e., the variability of stock prices due to a number of known and unknown factors), results in a much more conservative performance. In fact, according to the model, there is an almost 10% chance that the return on this portfolio will be negative.

So, can we pick a portfolio that is better? Well, what if we just invested an equal amount in each stock? We can simulate this portfolio with a 1/6 = 16.66% investment of the budget in each stock. The expected value of the monthly return based on point estimates is 0.63%; however, the simulation results in an average monthly return of 0.97%, with a standard deviation of 0.49%. Although the expected return is slightly lower than for the previous portfolio, the standard deviation has been greatly reduced, so this can be considered a "safer" portfolio. In fact, this portfolio has only a 1.55% chance of resulting in a negative return.

The question is: How can we find the best portfolio given the uncertainty in stock prices? We quickly realize that it is extremely unlikely that we will find it by "trying out" different portfolio alternatives by hand, since there are so many possible combinations of assets and budget allocations. So we need something more powerful to help us.

The solution to this challenge is to combine the advantages of optimization and simulation into a single framework. On the one hand, optimization can help us search for the best portfolio; on the other hand, simulation can ensure that we are not ignoring the uncertainty in stock prices.

A SIMULATION OPTIMIZATION SOLUTION APPROACH

As we learned in the previous section, simulation provides a way to evaluate the impact of changes to parameters and decisions in a model environment through the creation of "what-if" scenarios. Simulation also enables examination and testing of decisions prior to actually

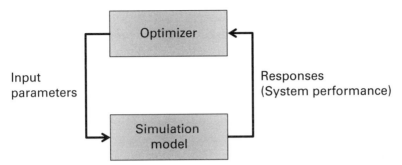

Figure 3.2   Black box approach to simulation optimization.

implementing them, thus saving the decision-maker's time and money. But perhaps most important, simulation enables the examination and evaluation of uncertainty and risks associated with a set of decisions, so that such risks can be understood and mitigated.

Although simulation provides all of these advantages, in most real-world situations—even one as simple as our portfolio example—the set of possible decisions is too large to enumerate, let alone to be searched through efficiently to find the best possible solution. We need a way to guide our search.

The merging of simulation modeling with optimization technology has provided the advance making it possible to solve this problem (April et al., 2004).

Once a simulation model has been developed to represent a system or process, we want to find a configuration that is *best* according to some performance criteria. In our portfolio example, this corresponds to finding investment levels in the set of six candidate stocks that will result in maximum expected returns at some minimum risk level. When a system is complex and the configuration depends on a number of strategic choices, the trial-and-error approach results in very limited success. In these cases, we use an optimization tool to guide the search for the best configuration.

In our approach, we view the simulation model as a "black box," meaning that we are only interested in obtaining an evaluation of performance from the simulation.

Figure 3.2 shows this black-box approach to simulation optimization. In this approach, the optimization procedure, called the *optimizer*, first chooses a set of values for the input parameters, also called the *decision variables*; next, the simulation model is run with these parameter values, and the optimizer obtains an *evaluation* of the performance of the system. The optimizer then uses this response from the simulation model to select the next set of values for the input parameters. This loop continues until certain stopping rules are met.

The key to a good simulation optimization approach is the search algorithm embedded in the optimizer, which makes decisions about the selection of new input parameter values at each iteration in the process.

Going back to our portfolio selection example, we now view our simulation model as a black box that provides a response about the expected return and standard deviation of returns of a selected portfolio. We wrap around it an optimizer that will guide us in selecting the allocation of our budget to each stock in order to obtain the maximum expected return at some minimum level of risk.

In order to do this, we need to tell the optimizer exactly what we want to achieve. Therefore, we first state our primary objective: *maximize the expected portfolio return*. If we run the optimization with this objective alone, we obtain the following portfolio recommendation:

- Invest the entire budget in ORACLE stock.

This is expected to produce the following results:

- Average monthly portfolio return = 1.06%
- Standard deviation of returns = 4.08%

In this case, since we did not tell the optimizer anything about the risk we are willing to accept, the optimizer naturally chose to invest the entire budget in the stock with the highest expected return, regardless of its volatility. We want to do better than this. So we amend our objective to include the additional goal: keep risk under a certain acceptable threshold.

We now express our desires as follows:

Maximize the expected portfolio return, but make sure the standard deviation of returns is no higher than 0.2%.

If we optimize this model, the best solution found—at iteration 861—recommends the following:

- 11% of budget should be invested in SUN MICROSYSTEMS stock;
- 46% of budget should be invested in MICROSOFT stock;

- 28% of budget should be invested in INTEL stock;
- 15% of budget should be invested in CISCO stock.

This yields the following results:

- Average monthly portfolio return = 0.96%
- Standard deviation of returns = 0.19%

It is notable that the optimal portfolio in this case includes NO investment in ORACLE stock despite it having the highest historical return. This is probably because ORACLE has the highest ratio of standard deviation to average return, making it the most volatile of the stocks, and thus increasing the riskiness of the portfolio. In fact, according to our amended model, the optimal portfolio described earlier has only a 0.04% chance of resulting in a negative return, which is a vast improvement.

SIMULATION OPTIMIZATION APPLICATIONS IN OTHER REAL-WORLD SETTINGS

The advantages of simulation optimization can not only be realized in financial modeling settings. A wide array of fields exists where systems exhibit high complexity and outcomes are sensitive to uncertainty. Examples are business processes, national defense systems, workforce planning, and so forth. When changes are proposed to a system in order to improve performance, the projected improvements can be simulated and optimized artificially. The sensitivity of performance objectives to proposed changes can be examined and quantified, reducing the risk of actual implementation, and increasing the confidence in the selected decision strategy.

In business process management, changes may entail adding, deleting, and modifying processes, process times, resource require- ments, schedules, work rates, skill levels, and budgets, making this a very fertile for such approaches. Performance objectives may include throughput, costs, inventories, cycle times, resource and capital utiliza- tion, start-up and set-up times, cash flow and waste. In the context of business process management and improvement, simulation can be thought of as a way to understand and communicate the uncertainty related to making the changes, while optimization provides the way to manage that uncertainty.

We now focus on two examples to showcase the application and advantages of a simulation optimization approach to business process management.

Selecting the Best Configuration in a Hospital Emergency Room

The following example is based on a model of a real emergency room (ER) process provided by CACI, and simulated on SIMPROCESS. Consider the operation of an emergency room in a hospital. Figure 3.3 shows a high-level view of the overall process. The process begins when a patient arrives through the doors of the ER, and ends when a patient is either released from the ER or admitted into the hospital for further treatment. Upon arrival, patients sign in, are assessed in terms of their condition, and are transferred to an ER room. Depending on their condition, patients must then go through the registration and treatment processes before being released or admitted into the hospital.

Patients arrive either on their own or in an ambulance, according to some arrival process. Arriving patients are classified into different levels, based on their condition, with Level 1 patients being more critical than Level 2 and Level 3.

Level 1 patients are taken to an ER room immediately upon arrival. Once in the room, they undergo their treatment. Finally, they complete the registration documentation process before being either released or admitted into the hospital for further treatment.

Level 2 and Level 3 patients must first sign in with an *Administrative Clerk*. After signing in, their condition is assessed by a *Triage Nurse*, and then they are taken to an ER room. Once in the room, Level 2 and Level 3 patients must first complete their registration documents, then go on to receive their treatment, and, finally, they are either released or admitted into the hospital for further treatment.

The ER treatment process consists of the following activities:

1. A secondary assessment performed by a nurse and a physician;

2. Laboratory tests, if necessary, performed by a patient care technician (PCT);

3. The treatment itself, performed by a nurse and a physician.

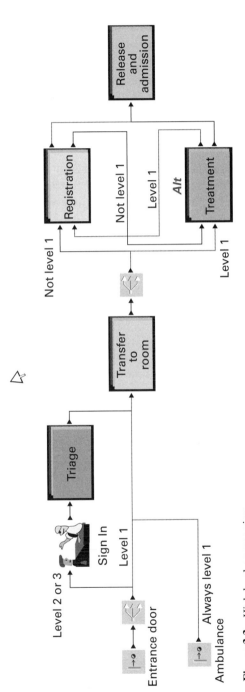

Figure 3.3 High-level process view.

The registration process consists of the following activities:

1. A data collection activity performed by an Administrative Clerk;
2. An additional data collection activity performed by an Administrative Clerk, in case the patient has Worker's Compensation Insurance;
3. A printing of the patient's medical chart for future reference, also performed by an Administrative Clerk.

Finally, 90% of all patients are released from the ER, while the remaining 10% are admitted into the hospital for further treatment. The final release/hospital admission process consists of the following activities:

1. In case of release, either a nurse or a PCT fills out the release papers (based on availability).
2. In case of admission into the hospital, an Administrative Clerk fills out the patient's admission papers. The patient must then wait for a hospital bed to become available. The time until a bed is available is handled by an empirical probability distribution. Finally, the patient is transferred to the hospital bed.

The ER has the following resources:

- Nurses
- Physicians
- PCTs
- Administrative Clerks
- ER rooms

In addition, the ER has one Triage Nurse and one Charge Nurse at all times.

Due to cost and layout considerations, hospital administrators have determined that the staffing level must not exceed seven nurses, three physicians, four PCTs, and four Administrative Clerks. Further, the ER has 20 rooms available; however, using fewer rooms would be beneficial, since the additional space could be used more profitably by other departments in the hospital. The hospital wants to find the configuration of the above resources that minimizes the total asset cost.

The asset cost includes the staff's hourly wages and the fixed cost of each ER room used. We must also make sure that, on average, Level 1 patients do not spend more than 2.4 hours in the ER. This can be formulated as an optimization problem, as follows:

Minimize the expected **total asset cost** subject to the following constraints:

- Average Level 1 cycle time is less than or equal to **2.4 hours.**
- Nurses are greater than or equal to 1 and less than or equal to **7.**
- Physicians are greater than or equal to 1 and less than or equal to **3.**
- PCTs are greater than or equal to 1 and less than or equal to **4.**
- Administrative Clerks are greater than or equal to 1 and less than or equal to **4.**
- ER rooms are greater than or equal to 1 and less than or equal to **20.**

This is a relatively simple problem in terms of size: six variables and six constraints. However, if we were to rely solely on simulation to solve this problem, even after the hospital administrators have narrowed down our choices to the above limits, we would have to perform $7 \times 3 \times 4 \times 4 \times 20 = 6720$ experiments. If we want a sample size of, say, at least 30 runs per trial solution in order to obtain the desired level of precision, then each experiment would take about 2 minutes.[1] This means that a complete enumeration of all possible solutions would take approximately 13,400 minutes, or about 28 working days. This is obviously too long a duration for finding a solution.

In order to solve this problem in a reasonable amount of time, we used the OptQuest® optimization technology integrated with SIMPROCESS (see www.OptTek.com for more information about the OptQuest optimizer). As a base case, we decided to use the upper resource limits provided by hospital administrators, to get a reasonably good initial solution. This configuration yielded an expected total asset cost of $36,840, and a Level 1 patient cycle time of 1.91 hours.

Once we set up the problem in OptQuest, we ran it for 100 iterations (experiments), and 5 runs per iteration (each run simulates 5 days of the ER operation). Given these parameters, the best solution found at iteration 21 was the following:

Nurses	Physicians	PCTs	Administrative Clerks	ER rooms
4	2	3	3	12

[1] We timed one experiment with 30 runs on a Dell Dimension 8100, with an Intel Pentium 4 processor @ 1700 MHz.

The expected total asset cost for this configuration came out to $25,250 (a 31% improvement over the base case), and the average Level 1 patient cycle time was 2.17 hours. However, looking at the probability distribution for cycle time, we see that there is still a 45% chance that the Level 1 patient cycle time will be greater than 2.40 hours.

After obtaining this solution, we redesigned some features of the current model to improve the cycle time of Level 1 patients even further. In the proposed model, we assume that Level 1 patients can go through the treatment process and the registration process in parallel. That is, we assume that while the patient is undergoing treatment, the registration process is being done by a surrogate or whoever is accompanying the patient. If the patient's condition is very critical, then someone else can provide the registration data; however, if the patient's condition allows it, then the patient can provide the registration data during treatment. Figure 3.4 shows the model with this change. By implementing this change in the optimized model, we now obtain an average Level 1 patient cycle time of 1.98 (a 12% improvement), with only a 30% chance that the Level 1 patient cycle time will be greater than 2.40 hours.

Upon re-optimizing this new model, given the change that we implemented, we obtain a new optimal solution in 28 iterations as follows:

Nurses	Physicians	PCTs	Administrative Clerks	ER rooms
4	2	2	2	9

This configuration yields an expected total asset cost of $24,574, and an average Level 1 patient cycle time of 1.94 hours, with less than a 5% probability that the cycle time will exceed 2.40 hours. By using optimization, we were able to find a very high-quality solution in less than 30 minutes, which provides a very acceptable level of risk in terms of service quality (i.e., cycle time). In addition, we were able to make changes to improve the model and re-optimize to find a better configuration. It is highly unlikely that this solution would be found relying solely on simulation.

Selecting the Best Staffing Level for a Personal Claims Process at an Insurance Company

The following example is based on a model provided by SIMUL8 Corporation. We used the *SIMUL8* simulation tool for the simulation, and *OptQuest for SIMUL8* for the optimization.

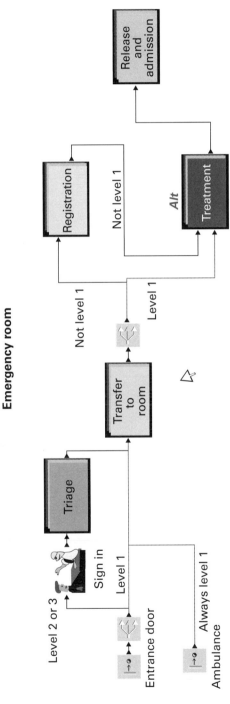

Figure 3.4 Proposed process.

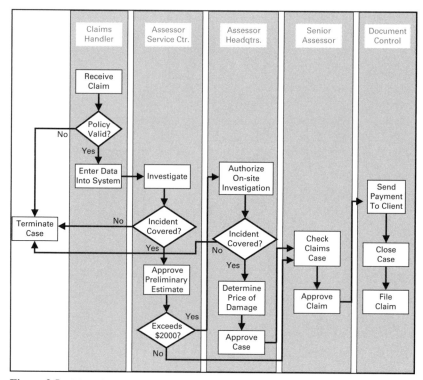

Figure 3.5   Map of personal claims process.

A personal claims department in an insurance company handles claims made by their clients. Claims arrive according to a Poisson process, with a mean inter-arrival time of 5 minutes. Figure 3.5 is a process map that depicts the personal claims process in terms of swim lanes.

The first lane corresponds to work done by a claims handler (CH) located at the client's local service center. Upon arrival of a claim, an assessor determines if the client has a valid policy. If not (5% of all cases), then the case is terminated; otherwise (95% of all cases), the assessor enters the appropriate information in the system.

In the second lane, an assessor located at the service center (ASC) receives the information from the claims handler. The assessor first determines if the claim is covered by the client's policy. If not (5% of all cases), the case is terminated; otherwise (95% of all cases), the assessor approves the preliminary estimate of the damage. If the damage exceeds $2000 (35% of all cases), the claim is sent to an assessor at headquarters for approval; otherwise (65% of all cases), it is sent directly to a senior assessor (SA).

Lane 3 corresponds to the assessor at headquarters (AHQ). The assessor first authorizes the on-site investigation of the accident. If the investigation determines that the incident is not covered by the client's policy (2% of all cases), then the case is terminated; otherwise (98% of all cases), a final price is determined and the case is approved.

In lane 4, the SA receives the claim, checks it for completeness, and provides the final approval. Once the claim is approved, it is sent to document control (DC).

DC, in lane 5, is in charge of processing the payment to the client, closing the case and, finally, filing the claim.

The objective here is to find staffing levels for each of the five resource types, in order to minimize headcount, while keeping average throughput above 1500 claims during 4 weeks. Each resource type has a maximum limit of 20 people, and the overall headcount in the process cannot exceed 90. The formulation of the optimization problem is as follows:

Minimize the **Headcount** Subject to the following constraints:

- Average Throughput is equal to or greater than **1500**.
- Claims handlers are greater than or equal to 1 and less than or equal to **20**.
- Service center assessors are greater than or equal to 1 and less than or equal to **20**.
- Headquarter assessors are greater than or equal to 1 and less than or equal to **20**.
- SAs are greater than or equal to 1 and less than or equal to **20**.
- DCs are greater than or equal to 1 and less than or equal to **20**.
- **The overall headcount cannot exceed 90.**

Once again, a what-if analysis of all the possible solutions to this problem would require examining and evaluating an unmanageably large number of scenarios—in this case, about 800,000. Optimization is necessary to find a good solution efficiently. A good starting point can probably be established by consulting with experienced managers in the insurance claims area, based on the expected demand of claims. We use OptQuest to optimize the staffing levels of this system. We run OptQuest

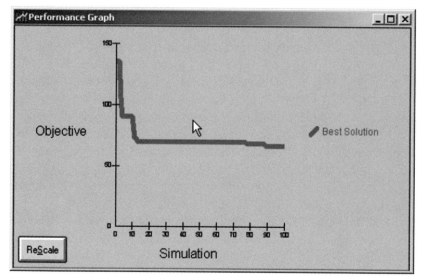

Figure 3.6 Performance graph for the optimization of the personal claims process.

for 100 iterations and 5 simulation runs per iteration. Table 3.2 shows a summary of the results, and Figure 3.6 shows the graph depicting the search of *OptQuest for Simul8* toward improving solutions.

The performance graph shows the value of the objective (in this case, headcount) on the y-axis, and the number of iterations on the x-axis. The performance curve (indicated in green) shows only improving solutions.

Since some of the solutions obtained from our optimization are relatively close in terms of throughput and cycle time, an analyst may want to reevaluate a set of the best *n* solutions to assess the precision of the results. In Table 3.2, we present the best 5 solutions obtained from our OptQuest run, by conducting an experiment of 20 trials for each solution. The information can now be given to a process manager. The manager can analyze the trade-offs between headcount and throughput or cycle time, to decide which configuration best aligns with service levels and process goals.

For example, it can be noted that solutions 1–3 are statistically the same. Solutions 4 and 5 are significantly better than 1–3 in terms of headcount, throughput, and cycle time; so, the analyst should pick one of these. Which one is better? We re-ran 60 trials for each of these two solutions, to obtain a 95% confidence interval for each of these

Table 3.2 Summary results

Simul8 results
Personal claims process—20 trials

| Solution | Claims handler | Assessor (SC) | Assessor (HQ) | Senior assessor | Document control | Throughput (claims) | | | Headcount | Average cycle time (min) |
						Lower Specification Limit	Average	Upper Specification limit		
1	9	17	17	15	16	1563.00	1568.00	1573.00	74	639.00
2	9	17	17	14	16	1559.00	1564.00	1570.00	73	658.00
3	8	17	16	15	16	1562.00	1567.00	1573.00	72	646.00
4	9	18	12	15	11	1611.00	1622.00	1633.00	65	503.00
5	9	18	11	15	11	1610.00	1621.00	1632.00	64	510.00

measures. In both cases, the confidence intervals overlap. In fact, the resulting confidence intervals for throughput are almost identical; the intervals for cycle time are also very close, with the expected cycle time for Solution 4 (503 minutes) about 1.4% lower than that for Solution 5 (510 minutes). The analyst should consider if the savings obtained from having one fewer assessor justifies such a small difference in cycle time. If so, then solution 5 should be chosen.

CONCLUSIONS

Practically every real-world situation involves uncertainty and risk, creating a need for optimization methods that can handle uncertainty in model data and input parameters. The combination of two popular methods, *optimization* and *simulation*, has made it possible to overcome limitations of classical optimization approaches for dealing with uncertainty, where the goal is to find high-quality solutions that are feasible under as many scenarios as possible. Classical methods by themselves are unable to handle problems involving moderately large numbers of decision variables and constraints, or involving significant degrees of uncertainty and complexity. In these cases, simulation optimization is becoming the method of choice.

The combination of simulation and optimization affords all the flexibility of the simulation engine in terms of defining a variety of performance measures as desired by the decision maker. In addition, as we demonstrate through illustrative examples in project portfolio selection, emergency room operation and insurance claims staffing, modern optimization engines can effectively enforce requirements on one or more outputs from the simulation. Finally, simulation optimization produces results that can be conveyed and grasped in an intuitive manner, providing an especially useful tool for identifying improved business decisions under risk and uncertainty.

REFERENCES

J. April, M. Better, F. Glover, and J. P. Kelly, "New advances and applications for marrying simulation and optimization," in *Proceedings of the 2004 Winter Simulation Conference*, December 5–8, 2004, Washington, DC, IEEE, 2004.

D.B. Hertz, "Risk analysis in capital investment," *Harvard Business Review*, vol. 57, no. 5, pp. 169–181, 1979.

H. Markowitz, "Portfolio Selection," *Journal of Finance,* vol. 7, no. 1, pp. 77–91, 1952.

W.F. Sharpe, "Capital asset prices: A theory of market equilibrium under conditions of risk," *Journal of Finance,* vol. 19, no. 3, pp. 425–442, 1964.

CHAPTER 4

Optimal Learning in Business Decisions

Ilya O. Ryzhov

Robert H. Smith School of Business, University of Maryland, College Park, MD, USA

INTRODUCTION

We know that business decisions are made under uncertainty. The demand for a product at a retail store varies from one week to the next, and the exact weekly sales cannot be known in advance. Customer response to a new product or service, offered, for example, through a website, is similarly uncertain. Even if a reliable forecast of sales is available, such a forecast will typically model some sort of average or aggregate behavior across a population of customers. Any individual customer is unlikely to behave exactly according to the forecast. Even the supply of the product may be uncertain. For example, a small electricity operator may generate electricity from a wind farm and sell it back to the grid; the firm's revenue thus depends on volatile wind speeds. A manufacturer may contract with suppliers that experience occasional shortages.

Management science and business analytics offer many ways to deal with uncertainty. Chapter 3 showed how simulation–optimization can be deployed to optimize decisions when the causal relation between

Breakthroughs in Decision Science and Risk Analysis, First Edition.
Edited by Louis Anthony Cox, Jr.
© 2015 John Wiley & Sons, Inc. Published 2015 by John Wiley & Sons, Inc.

controllable inputs and probabilities of outcomes can be quantified and simulated. Sashihara (2011) describes many case applications in such business problems as logistics, pricing, marketing, and new product design, combining (i) statistical forecasts of the future, based on historical data collected from the field; (ii) rigorous models for decision-making; and (iii) probabilistic estimates of the likelihood of different future scenarios. In this chapter, we focus on a different point, namely, that these decisions also involve the additional dimension of "uncertainty about uncertainty," or *environmental uncertainty*. Simply put, although we may develop models that consider uncertainty in customer demand or product supply, we have no way of knowing whether these models are completely accurate. In fact, most models have some degree of inaccuracy. The underlying business processes may also change over time; our forecast of demand may change completely as new data come in. Our belief about the "average demand" or "best price" for a product will be repeatedly adjusted to reflect new information. However, high-impact decisions often have to be made with a limited amount of information available. For example, a company seeking to do business with suppliers from developing economies faces a considerable amount of uncertainty about the suppliers' reliability, often without a great deal of historical data available. We may wish to consider the likelihood of a shortage as a factor in our decision to sign a contract with such a supplier, but we do not even know what the *likelihood* is, to say nothing of when the shortage might occur.

A manager making decisions in an uncertain environment faces a double challenge. First, the decision should consider multiple future scenarios and account for the possibility of unplanned shortages, downtime, fluctuations in demand, and so forth. This aspect can be considered by applying widely used analytical techniques, such as forecasts or models based on historical data. Then, once such models have been developed, the decision should be additionally adjusted to account for the chance that *the model itself* may be wrong or, alternately, that the model may itself be *subject to change* over time. If we have a way to anticipate such changes before they occur, our decisions will come closer to optimizing the real-world process, rather than just our current model of that process. The field of *optimal learning* studies exactly this challenge: how to move from *experience-based* decisions, which reflect the sum total of our current knowledge about a problem, to *anticipatory* decisions that consider the potential for error or change in that knowledge.

Anticipatory decisions are entering current business practice. A good example is e-commerce, where new information about customers is collected around the clock. When you shop online, you may notice that the price of a product changes frequently, sometimes multiple times per day, even for low-volume products (such as textbooks) that may spend a long time in inventory. New orders are not arriving often enough to change the demand forecast or the projected revenue curve so quickly or drastically. Rather, the changing prices reflect the retailer's uncertainty about customer response. Demand is not necessarily expected to spike after a price drop, but the retailer chooses to *experiment* with low prices, simply to learn more about customer reactions. Likewise, the retailer may nudge prices upward—historical data may suggest that demand will decrease, but the retailer would like to obtain more precise information about exactly which prices customers are willing to accept.

The retailer does not necessarily expect to improve sales immediately as a result of this type of experimentation. We may try a higher price, only to find that demand falls even more than we predicted. Alternately, lower prices may not increase sales by enough to compensate for the lost revenue. However, the results of these experiments feed back into the retailer's databases and models, increase the precision of future forecasts, and generally make it possible to make more informed pricing decisions that lead to greater revenues later—eventually, enough to make up for the cost of experimentation (relatively low in e-commerce, due to the ease of implementing a price change). The key idea is that *information possesses inherent economic value*. More information now leads to better decisions later. This long-term improvement can be used to valuate the experiment and determine whether it was worth the cost. The essence of anticipatory decision-making is to gauge this improvement before the experiment is conducted and make an informed trade-off between immediate economic benefit (lost revenues now) and long-term performance (improved revenues later).

Once we acknowledge that (i) any model or forecast is inaccurate and (ii) new information is valuable for its potential to improve the model, it is easy to think of many other situations where environmental uncertainty plays an important role. Below are some additional examples of high-impact decisions subject to uncertainty about uncertainty:

- *Marketing.* The field of marketing analytics studies the extraction of information from customer data. By relating customer response to different design attributes of marketing campaigns (e.g., the medium used, whether mail, phone, or online; the format of a piece of mail or the length of a phone call; the use of particular images, slogans, or text in an advertisement), we can predict how well a particular advertising campaign will do with different segments of the customer base. Such forecasts can be used to guide the design of the next campaign. However, even large amounts of data can produce ambiguous or contradictory forecasts. To obtain more precise information about a customer segment, a company may test a particular design attribute on a small portion of the customer population, before launching the campaign.

- *New product design.* The profit margin and demand for a new product is unknown, so it is unclear how many units should be produced for a first run or which design should be selected from among a few competing alternatives. However, we can run test markets on a few designs to get a better sense of the customer response. Each test market costs both time and money, so we may not be able to test every design or even to get a precise estimate for the designs that we do test. Nonetheless, if properly used, the test markets can improve the performance of the product upon launch.

- *Global operations.* The performance of a supply chain is subject to uncertainty about the reliability of suppliers in developing economies (Wang, Gilland, and Tomlin, 2010). By signing a contract with one supplier, we can learn more about the supplier's service reliability, but the costs of an uninformed decision can be high. Still, trying one supplier may also provide information about other suppliers in the same region or offering similar products.

- *Calibrating a design.* The operations of a fleet of aircraft (flight schedules, repair frequencies, customer demands, etc.) can be modeled using a computer simulation. The simulation outputs statistics such as the amount of unsatisfied demand or the time spent on the ground. These statistics are affected by different "rule-based" decision-making strategies, such as a penalty or opportunity cost for delayed flights or a cap on the number of aircraft to keep at a single airport. By experimenting with these different settings, a better rule can be found.

- *Clinical trials.* In clinical trials (or any other setting where new and risky technologies are researched), it is necessary to screen out a large number of unpromising experimental drug compounds and identify a small shortlist of candidates. Then, it is necessary to conduct

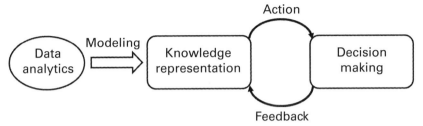

Figure 4.1   Process diagram for anticipatory decision-making.

additional experiments on the most promising compounds to eventually select the best, which is put into production. The cost of a single experiment on one candidate may be quite high; for example, it may be necessary to monitor the health condition of a group of mice for a week. It is impossible to test every possible compound, which makes it important to allocate the experiments wisely in order to maximize the chances of finding the best (Negoescu, Frazier, and Powell, 2011).

- *Operational flexibility.* An industrial facility has high electricity consumption. During a spike in the price of electricity, it is actually beneficial to ramp down operations (ideally to stop altogether) to reduce energy costs until the spike passes. Ramping up takes time, so we have to predict the spike before it occurs. If our prediction is correct, the cost savings will be significant; if not, we will waste electricity in ramping back up as well as delay production. Energy prices are highly volatile, and existing price models and forecasts often have low accuracy.

- *Hiring and retention.* Suppose that we are hiring a large number of workers for a manufacturing or service operation. As workers master various tasks, their performance tends to improve over time. If a new worker's performance is lower than that of other workers, it may be that the worker is insufficiently qualified for the job, but it may also be that the worker simply needs more time to improve. Firing the employee means that the firm will incur additional costs for hiring and training a replacement, so the decision to fire or retain must consider the worker's potential for future improvement (Arlotto, Chick, and Gans, 2010).

Figure 4.1 illustrates the relationship between modeling, decision-making, and information. The first step toward anticipatory decision-making is to develop a model for *knowledge representation*. We may also refer to this model by other names such as "state of knowledge," "belief model," or simply "beliefs." As the name implies, this model encompasses our current understanding of the problem we are solving, based on the most recent historical data available. A substantial amount of prior work goes into designing this model, represented by "Data analytics" in the diagram. For example, in a marketing application, this step would consist of mining historical data, identifying customer segments, screening out unimportant campaign and customer attributes, and using statistical regression or other methods to estimate the effects of important attributes. Inherent in the belief model should be some sort of "margin for error," a measure of how likely the estimated effects are to be inaccurate.

The next step is to use the knowledge representation to take an *action*, that is, to use our current understanding of the problem to make a decision. It is important for the knowledge representation to be as compact as possible. We need a simple way to express our beliefs about the problem, so that we may translate them to a rule for making decisions or interface them with optimization software to obtain a recommendation for the next decision. Either way, once the action is taken, the problem briefly passes out of our control, and we observe *feedback* from the field. For example, if the action is to launch an advertising campaign, the feedback is a measure of customer response; if the action is to adjust a price, the feedback is the resulting sales. The feedback need not be precise. We do not necessarily know if sales dropped because our price was too high or simply due to random fluctuations in the number of people shopping for the product on that day. All we can observe are the sales figures themselves.

Finally, once we receive the feedback, we adjust our beliefs about the problem to account for it. If a higher price led to a larger drop in sales than our model predicted, the model should be adjusted to make the forecast less optimistic. This modified or improved model can then be used to guide the next decision, leading to a constant loop between beliefs, decisions, and feedback. This loop implicitly assumes that we have a way to efficiently update the model to account for new knowledge. In this chapter, we will discuss two such "learning models" that are especially easy to update with new information. In general, it is important to make sure that beliefs can be represented by a small set of numbers (such as means and standard deviations of forecasts) that can then be easily updated.

So far, we have described experience-based decision-making, where we learn from our mistakes and continually update our beliefs with new information. The key to anticipatory decision-making is to acknowledge and account for the potential feedback *before* taking the next *action*. Our next decision should not only depend on the current beliefs but also on the potential of the decision to change those future beliefs. To put it a different way, our decision should look ahead to the future, more accurate belief model that will come about as a result of the decision. We should not only maximize the benefit of the next decision but also work to improve the model and put ourselves into a more advantageous position for the next round.

In the rest of this chapter, we will explore two examples where anticipatory decisions offer significant improvements over experience-based decisions. First, we discuss a newsvendor problem, a simple example from revenue management where the demand for a product is subject to environmental uncertainty. We show how this environmental uncertainty can be compactly represented and updated and then leveraged to improve long-term profits. Next, we consider the "selection" problem, where the goal is to identify the best among a fairly small set of decisions, such as supply contracts, operational policies, or R&D projects. In this problem, the cost of implementing an alternative (such as developing a research project and putting it into production) is high, making it important to collect more information before committing to a final decision. Again, we show how our uncertainty about the performance of different alternatives can be represented and updated with feedback from experiments (whether in the field or using a computer simulation). Then, we show how, in this setting, the *potential* of an experiment to improve our implementation decision can be quantified and used to determine how the experimental budget should be spent. Finally, we apply this method to the problem of selecting the best rule-based operational policy for managing inventory.

While the mathematical models in these examples will not necessarily apply to every business problem, they do offer some general insights that also hold for other models. Most important is the idea that *uncertainty can be valuable*! Although uncertainty limits our ability to maximize profits or minimize costs, it also offers the potential for improvement. A decision that seems to offer poor performance, but that exhibits a high degree of uncertainty is more likely to be better than we think. It may even be better than what seems to be the best decision at the moment—we may only need to invest a portion of our information budget to make sure.

The following sections include occasional references to other reading, much of it substantially more technical than this chapter. Readers interested in the big picture should feel free to skip these references, though some of them consider much more complex learning problems than those discussed here. Powell and Ryzhov (2012) present a more technical overview of the field.

OPTIMAL LEARNING IN THE NEWSVENDOR PROBLEM

Our first example of environmental uncertainty in decision-making will consider the newsvendor problem, a classic model in operations management (Stevenson, 2008). Imagine a small firm selling a single product. A manager must decide how many units of the product should be produced; then, a random demand is observed and the firm sells as many units as possible. If the demand is higher than the production quantity, the entire supply can be sold, but there is an opportunity cost since profits could have been higher if we had stocked enough to satisfy all the demand. If the demand is lower than the production quantity, the firm is able to satisfy every customer but is left with extra inventory. For simplicity and to emphasize the trade-offs involved, we suppose that extra inventory is unsalvageable.[1] Because demand is random, it is impossible to predict just the right production quantity for every situation. However, the manager can make a decision that will work well on average.

Suppose that the marginal production cost of the product is c dollars per unit, while the selling price is p. The quantity chosen by the manager is represented by q, while D denotes the random demand. If we observe $D=d$, the profit realized by the firm is given by

$$\pi(q, d) = p \min(q, d) - cq. \tag{4.1}$$

---

[1] The name "newsvendor" comes from an analogy to a newsstand—leftover copies of today's paper are essentially worthless and cannot be stored for sale at a future date.

The quantity $\min(q,d)$ represents the most that we can sell: either we stocked less than the demand (and sold all of it), or the demand was smaller than we anticipated, and we ended up with extra inventory.

Of course, (4.1) will give a different result depending on the observed value d, so this equation is not useful for actually choosing an order quantity. However, we can calculate the *expected* profit realized by the firm as an average over all possible values d of D. We write this as

$$\bar{\pi}(q) = pE\big[\min(q,D)\big] - cq, \tag{4.2}$$

a quantity that can be calculated exactly if we make some assumptions about the probability distribution of D. To give a single example, suppose that D follows an exponential distribution with rate r. Recall that the exponential distribution is used to model random variables whose values are positive (as is the case for demand) and continuous (which can happen if fractional units of the product are possible, such as 2.3 pounds of sugar). The solid line in Figure 4.2a shows the probability density of an exponential distribution with rate $r=2$. Values with higher density are more likely to be observed; under this distribution, the random variable tends to take on small values, but very large values can also be observed occasionally. The average or expected value of D is $E(D)=1/r$, so larger rates (somewhat counterintuitively) lead to smaller demands.

By integrating (4.2) over the exponential probability density, we obtain

$$\bar{\pi}(q) = \frac{p}{r}\big(1 - e^{-rq}\big) - cq, \tag{4.3}$$

an expression that only depends on the chosen production quantity. Again, while the actual realized profit will depend on the exact demand that we will ultimately observe, (4.3) represents a forecast or an expectation of the profit for a given production quantity q. The "randomness" is removed from this expression because we are considering an average over all possible demands. After some more calculus, we can find

$$q^* = \frac{1}{r}\log\left(\frac{p}{c}\right), \tag{4.4}$$

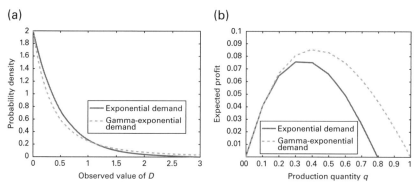

Figure 4.2 Comparison of (a) probability densities and (b) expected profit curves for newsvendor models under exponential and gamma-exponential demand assumptions.

the production quantity that *maximizes* the profit function in (4.3). While this quantity is not guaranteed to perform well in every scenario, a risk-neutral manager would choose this decision because it produces the best performance on average.

So where does the idea of learning come into play? Our analysis has made the assumption that D is exponentially distributed with rate r. Even supposing that we agree with the overall choice of an exponential distribution, it is unlikely that the value of r used in our calculation is actually the true rate. It is more likely that we are working with some sort of guess or estimate of the rate, possibly based on our prior experience. For example, suppose that we have some historical demand data $D_1, D_2, \ldots, D_k$ for k planning periods. A natural approach would be to use

$$\frac{1}{r} \approx \frac{D_1 + \cdots + D_k}{k}, \tag{4.5}$$

that is, to assume that the true average demand $1/r$ can be accurately represented by the sample average of the k previous demands that we have in our data. We could then plug (4.5) into (4.4) and arrive at

$$q^{\mathrm{PE}} = \frac{D_1 + \cdots + D_k}{k} \log\left(\frac{p}{c}\right), \tag{4.6}$$

where the notation PE stands for "point estimate," representing the fact that we are using a historical average as a direct stand-in for the true demand rate.

The main insight of optimal learning is that there is more uncertainty in this problem than there seems to be. We are uncertain about the random demand and therefore model the demand using a probability distribution. What is more, however, we are uncertain about the probability distribution itself. Rather than assume we know r, as we do in (4.5), we will now model r itself as a random variable. The "randomness" of r represents our uncertainty about the exact value of the rate: although we may believe that (4.5) is accurate, the exact value of r may be different. However, we should not discount the historical data entirely: we believe that the true value of r is more likely to be close to the prediction made in (4.5), although we may allow for the possibility that the historical data is very inaccurate. These concerns can be accommodated by choosing a good probability distribution for r. In this example, we will choose a gamma distribution with two parameters:

$$a = k,$$
$$b = D_1 + \cdots + D_k.$$

Roughly, the first parameter a represents the amount of prior data that we have, and the second parameter b represents the total demand observed across the data. The expected value of r, under the assumption of a gamma distribution, is

$$E(r) = \frac{a}{b} = \frac{k}{D_1 + \cdots + D_k},$$

which seems to be in line with (4.5). We still think that the reciprocal of the sample average computed from the data is the most likely value of r, but we allow the possibility of other scenarios.

Once we make r into a random variable, however, we have to revisit the distribution of the demand D. In effect, we are saying that if r is known, then D follows an exponential distribution (i.e., the demand is "conditionally exponential"). If we consider the uncertainty about the demand together with the uncertainty about the demand rate r, the distribution of D is modified. The dashed line in Figure 4.2a shows this modified density under the assumption that r follows a gamma distribution with $a = 2$, $b = 1$. Notice that, in this case, $E(r) = 2$, whereas the solid line assumes that $r = 2$. However, there is a difference between assuming that r is *exactly* equal to some value and assuming that r is equal to that value on *average*. By allowing r to be random, we see that larger values of the observed demand (1.5 and larger) are now slightly *more* likely to occur (have higher probability density) than they were when r was fixed.

Since the distribution of D has changed, the expected profit calculation is also different (and more difficult). Under the gamma assumptions on r, (4.3) changes to

$$\bar{\pi}(q) = p\frac{b}{a-1}\left[1-\left(\frac{b}{b+q}\right)^{a-1}\right] - cq, \tag{4.7}$$

and the profit-maximizing price becomes

$$q^{GD} = b\left[\left(\frac{p}{c}\right)^{\frac{1}{a}} - 1\right], \tag{4.8}$$

where GD stands for "gamma distribution," representing the fact that our model now includes uncertainty about r.

Figure 4.2b compares the expected profit curves for $r=2$ and r random with $E(r)=2$. When r is unknown, the maximum possible profits are expected to be higher since large demands are believed to be slightly more likely to occur. More importantly, the profit-maximizing production quantity is *greater* when r is random than it is when r is known. In fact, it can be shown that this is *always* true: the GD quantity with parameters a and b is always greater than the PE quantity that assumes $r=a/b$. This fact leads us to the first major insight from optimal learning: *uncertainty invites risk-taking.* Increasing the production quantity is risky; while we can achieve higher profits if the demand is high, there is also a higher chance that we will be stuck with worthless inventory. However, if the demand distribution is itself uncertain, there is an incentive to adjust the production quantity a bit higher than it would be in a certain world, because there is now a chance that the demand distribution could be *better than we think.*

But what is the role of *learning* in this process? So far, we have described how good decisions are affected by the amount of uncertainty experienced by the decision-maker about the problem. Learning occurs, not only when uncertainty has an impact on our decision, but when our decision has an impact on this uncertainty. This can happen when the firm wishes to maximize profit over multiple planning periods and the manager's beliefs about the demand rate are updated with new information. In other words, every time the manager chooses a production quantity and observes the subsequent demand, that observation will now change the assumed distribution of the demand rate r. The next production quantity will then be calculated from (4.8) using new, different a and b values. If we are able to observe the exact value of the demand, regardless of whether or not we had enough inventory to satisfy it, the update is straightforward: we simply increase a by 1 (reflecting the fact that we have one new piece of data) and add the demand value itself to b. In this case, it is enough to recompute (4.8) for each planning period using the most recent set of data—since the product cannot be salvaged at the end of a planning period, and since we observe the exact demand regardless of how much we produce, our production decision today will affect our profits for the current planning period, but will have no impact on future decisions.

However, it is much more realistic to suppose that we will not get to see the exact demand. Rather, we will observe our *sales*. If we sell everything we produce, we know that the actual demand may have been higher (i.e., we could have been able to sell more), but we do not know exactly how many potential customers we lost. By contrast, if we have leftover product at the end of the planning period, we know that our sales satisfied all the demand that there was. The difference between these two events is known as *censored information* and turns out to have a profound impact on decision-making. If we assume, again, that demand is exponentially distributed with rate r, and r itself follows a gamma distribution, with parameters a and b, we can use the following mechanism, created by Lariviere and Porteus (1999), to learn from censored information:

- Given the current information a and b, choose a production quantity q.
- If we have leftover inventory at the end of the planning period, increase a by 1 and add the amount sold to b.
- If the entire production quantity is sold, add this quantity q to b. Do not change a.

We do not delve into the mathematical justification for this approach, but we briefly describe the intuition. Recall that a represents the amount of data that we have, while b represents the total demand that has been observed. If we do not sell everything that we produced, we know that the demand for that planning period was exactly equal to our sales. We thus add the quantity sold (i.e., the demand) to b and increase a by 1 to indicate that we have obtained a new piece of data. However, if our entire stock is sold, we still add sales to b (as we have observed part of the demand, at least), but we do not count this observation as a "complete" piece of data and thus do not add it to a.

This seems straightforward, but introduces a serious complication into the manager's decision of choosing a production quantity. As we have observed before, large values of q are risky because they lead to a higher chance of having worthless leftover product. Now, however, leftover product is not exactly worthless—the more we produce, the more likely we are to get a complete observation of the demand. Such observations come at a high price, since we cannot sell the extra product, but they provide more accurate information about the demand distribution and thus can potentially help us make better, more informed decisions in future planning periods. In other words, the manager's decision does not only determine the firm's profits, but it also determines the quality of the information collected by the firm. It may therefore be beneficial to sacrifice some profit in the short term in order to improve the firm's production strategy over the long term.

Of course, the manager can always use the quantity provided by the GD formula in (4.8). However, even this strategy will no longer be optimal, because it does not explicitly consider the effect of today's decision on the information that will be available in the future. For example, all other things being equal, we should make different decisions early on in the planning horizon than at the end. Intuitively, we have more freedom to experiment early on—if we are unsuccessful, there will still be many opportunities to make up for it, and if we are successful and learn something useful, that information will benefit us for a long time to come. It goes without saying that day-to-day profits matter even when we are experimenting. An extremely poor result early on will be difficult to compensate, even if we make good decisions later.

In fact, Lariviere and Porteus (1999) provide a way to compute the trade-off between profits and information exactly. To find the best production quantity, the manager needs to know the total number of planning periods that should be considered. Let N denote this number. For every time period $n = 1, \ldots, N$, the production quantity is determined based on functions $B_n(a)$ and $R_n(a)$, computed as follows:

- For any a, $B_{N+1}(a) = R_{N+1}(a) = 0$.
- For $n = 1, \ldots, N$, we calculate:

$$B_n(a) = p + R_{n+1}(a+1) - R_{n+1}(a)$$

$$R_n(a) = p - c - ac\left[\left(\frac{B_n(a)}{c}\right)^{\frac{1}{a}} - 1\right] + R_{n+1}(a+1).$$

- The optimal production quantity at time n is given by

$$q_n^* = b\left[\left(\frac{B_n(a)}{c}\right)^{\frac{1}{a}} - 1\right],$$

where a and b represent the most recent information about r.

Calculating this quantity requires more work: in order to know the best decision at time n, we first need to calculate the best decisions for all future time periods for a variety of scenarios. However, the calculation is easy to automate and can be done in a spreadsheet for relatively small N. In our discussion, we focus on the implications. Figure 4.3a compares the optimal production quantities, in a problem with $N = 10$ planning periods, to those obtained from GD and PE, under the same set of information. This last detail is crucial—just to demonstrate the concept, we are assuming that the manager always holds the same information (represented by a and b), and we only vary the number of planning periods remaining. In other words, Figure 4.3a shows how the best possible decision changes if we have the same information early versus late in the planning horizon. The GD and PE production quantities do not change with the planning period, since both of those formulas only seek to maximize one-time profits with no regard for the total number of decisions we will have to make. The optimal strategy, however, recommends more production early on. Figure 4.3b shows the probability of shortage (i.e., the probability that demand will exceed

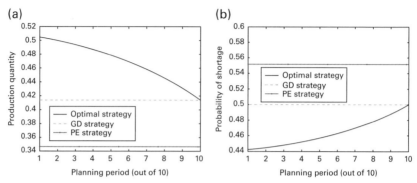

Figure 4.3   Comparison of (a) optimal production quantities and (b) shortage probabilities for different strategies in a problem with 10 planning periods under the same information.

production) for each planning period. If we follow the optimal strategy, our higher production will reduce the likelihood of a shortage and increase the chance that we will get a complete observation. As we move forward in the planning horizon, information becomes less valuable; when we make the final decision, information has no value at all (because it can no longer be used to improve future decisions), and it is optimal to follow GD.

Our discussion reveals an even stronger insight: not only does uncertainty invite risk-taking, but so does learning. All else being equal, it is beneficial to behave more aggressively early on in the planning horizon. This means that we will be more likely—6% more likely, in Figure 4.3b—to incur extra costs due to unsalvageable leftover product. At the same time, however, we will also be more likely to obtain valuable complete demand data that will give us a more accurate model for future decisions. As we approach the end of the planning horizon, we ramp down our aggressive production strategy and place more emphasis on maximizing here-and-now profits. All together, this trade-off will produce higher cumulative profits in the long run than a strategy that does not look beyond the current planning period.

From an analytical point of view, the key aspects of the model can be summarized as follows:

- Probability distributions play a double role in optimal learning. Not only do they represent the exogenous information we receive from the field, such as customer demand, but they can also be used to model our *beliefs* about that information. Thus, we model demand using one random variable and then place a separate distribution on r, a number that characterizes the average demand.

- Belief distributions add uncertainty. Not only do we have inherent uncertainty about the demand from one planning period to the next, we have additional uncertainty about the precise *distribution* of that demand. In effect, we have to consider many possible scenarios: in each one, the demand is random, but the average demand may be slightly higher or lower than our existing estimate.

- New information changes the belief distribution, a process that itself has inherent economic value. By increasing the production quantity, we are in some sense investing in better information that can provide us with a more accurate demand forecast one planning period later. This information can be valuated and traded off against short-term economic gains.

OPTIMAL LEARNING IN THE SELECTION PROBLEM

In the newsvendor problem, we collected information about a single unknown value, namely, the demand rate r. For our second example, we consider a problem where there are multiple unknown values, and we can choose which one we want to learn about. This problem is known as *ranking and selection* in the academic literature. In a business context, the problem simply consists of choosing a "design alternative," broadly defined, from a small set of candidates. Design alternatives represent high-impact managerial decisions such as the following:

- System designs such as facility layouts
- Long-term contracts with different partners or suppliers
- Commitments to build new facilities (factories, power plants, etc.) in certain locations
- Investments into new research or technology
- Hiring or retention policies
- Other decisions with significant economic or financial impact

Typically, the set of possible decisions will be small, because decision-making is occurring at a high level after many obviously inferior decisions have already been eliminated. The task of the manager or executive is to choose the best among the most promising decisions. In other words, we wish to identify the decision with the highest value, measured in terms of return (on an investment strategy or research program), cost (e.g., of a production or worker retention policy), or productivity (of a facility with a certain layout). The difficulty of this task comes from the fact that we do not know these values exactly. However, we do have access to some prior information about the values. For example, manufacturers can construct and use simulation models to evaluate the throughput of a factory with a particular layout (Eneyo and Pannirselvam, 1998). Simulation has also been used to evaluate the potential power output of a new wind farm at a candidate location (Marmidis, Lazarou, and Pyrgioti, 2008), a process that often requires sophisticated physics-based models of wind speed, wind turbine operation, and power conversion. In global operations, a belief about the reliability of a supplier in a developing economy can be formulated based on prior experience with suppliers from the same country or region. The potential profitability of a new product can be estimated by running one or more preliminary test markets. However, all of these methods will only give us an approximation of the true value of a decision.

Consider the following simple example with three decisions. For each one, we have an estimated value, obtained through simulation, historical data, field studies, or other means. We also have a standard deviation, which roughly represents a degree of confidence about the estimated value (lower standard deviation means that the estimate is more precise). For example, if we have more historical data about one particular decision, we will likely have a better understanding of how well it performs. Suppose that the three decisions are described by the following table.

Alternative	Estimated value	Standard deviation
1	$5M	$0.5M
2	$4.5M	$1M
3	$5.5M	$0.25M

If we had to make our final decision based purely on these numbers, we would probably prefer the third option: not only do we believe it to be the best (a risk-neutral decision-maker would trust the estimated values), we also are more confident in this belief than we are for the other two decisions. The real challenge appears when we can collect still more information before making the final decision. Suppose that one more simulation, test market, or field study can be conducted for any one of the three options. Our estimate for the alternative we choose will be adjusted based on the results (e.g., if alternative #3 performs worse than expected, the estimate of $5.5M can be reduced), and then we will make the final implementation decision. Which alternative should we learn about?

The answer is not at all clear. Should we spend our "information budget" on option #3, because it seems to be the most promising? In a way, this may be a waste: we already seem to know this alternative fairly well, and our experiment may simply tell us what we already know. By contrast, option #2 appears to underperform, but it also has the highest standard deviation, suggesting that there may be a sizable chance that this alternative is much better than we think. The "right" alternative to learn about should look reasonably good; if we strongly believe that, regardless of what its exact value is, there is no chance that it could ever be the best decision, there is no point in learning any more about it. At the same time, we should learn about an alternative with reasonably high uncertainty; otherwise, there is no reason to spend extra time collecting information. Our *learning decision* or *measurement decision* should thus depend on some combination of the alternatives' estimated values and standard deviations. For example, one simple way to measure the "potential" of an alternative is to simply add these two quantities. If v_i is the estimated value of alternative i and s_i is the standard deviation of the estimate, we can simply choose to learn about whichever alternative achieves the largest value of $v_i + s_i$. This will favor alternatives that seem to be good, but it will also favor alternatives with more uncertainty. Under this system, alternative #3 from our example still seems like the best option.

The issue here is that there is no clear reason why v_i and s_i should be equally important. A more conservative decision-maker would prefer to put more weight on the estimated values. A more exploratory strategy would put more weight on the uncertainty, that is, it would prefer to learn about alternatives that are not well understood. To account for these different priorities, we can modify our strategy as follows:

- Calculate the potential of an alternative as

$$u_i = w_i v_i + (1 - w_i) s_i,$$

   where the weight w_i is between 0 and 1. The closer w_i is to 1, the more conservative our decision will be.
- Learn more about the alternative with the largest potential.

In fact, this simple strategy (known in a slightly different form as "interval estimation," introduced by Kaelbling, 1993) can often be the right thing to do. The catch is that the weighting factor w_i has to be chosen carefully, often through trial and error. There appears to be no one value of w_i that will always work well; the "right" value depends on the estimated values themselves, the amount of uncertainty in our problem (if s_i is very large for every i, we may need to scale the weight up to make sure the estimated values are playing a part in the decision), and the precision of the information we are collecting. This last part is crucial: if the new information has the potential to significantly reduce our uncertainty, we may prefer an alternative with higher s_i, just to get the chance to estimate its value more accurately. However, if the new information is unreliable and exhibits a large amount of variation, this kind of experimentation will not be as valuable. The exact balance between all of these factors is difficult to express using the simple weighting rule described earlier.

To develop a more sophisticated approach, we need a clear sense of what we are looking for when we collect information. As we have discussed, we do not necessarily want to learn more about the alternative that seems to be the best (has the highest value of v_i). What we really want is to find an alternative that has high *potential*, not only to be the best, but to change our overall picture about the problem. We want to collect information that has high impact—that is, information that may cause us to change our plans for the implementation decision. As we shall see in the following, this can occur in two ways.

Figure 4.4 shows probability density curves for the three alternatives in our recurring example. Just as in the newsvendor problem, we

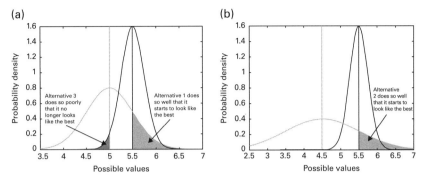

Figure 4.4   Evaluation of the potential of (a) alternative #1 and (b) alternative #2 against the current best option.

model the "true" value of an alternative as a random variable. The randomness here represents our uncertainty about the precise quality of the alternative. Thus, if alternative #3 is believed to have a value of $5.5M and the standard deviation of this estimate across our historical data is $0.25M, we can use a normal distribution with a mean of 5.5 and standard deviation of 0.25 to model our beliefs about #3. What we are saying here is the following:

- We think that the value of this alternative is $5.5M.
- The actual return from this alternative may be different, but we think that values closer to $5.5M are more likely.
- There is a chance that our estimate is completely wrong, and the actual return may be $1M or $10M, but this is less likely compared to values around $5.5M.

A normal distribution easily models all of these statements. The normal probability density assigns the most weight to values that are close to the mean (i.e., these values are considered to be the most likely). However, some weight is still assigned to values that are far away from the mean; this weight will be greater if the standard deviation is higher. Figure 4.4a shows the normal belief densities for alternatives #1 and #3, with estimated values of $5M and $5.5M. The center of the density for alternative #1 is smaller than for #3, but the curve overall is wider. In fact, belief density #1 assigns a substantial weight to values that are greater than 5.5, the center of the other curve. The exact value of this weight can be found by calculating the area of the blue region in

Figure 4.4a. Suppose that V_1 is normally distributed with a mean of 5 and standard deviation of 0.5, V_2 is normal with a mean of 4.5 and standard deviation of 1, and V_3 is normal with a mean of 5.5 and standard deviation of 0.25. The probability we are interested in is

$$P\left(V_1 \geq 5.5\right) = P\left(Z \geq \frac{5.5 - 5}{0.5}\right) = P\left(Z \leq -1\right) = 0.1587,$$

where Z is a standard normal (mean 0, variance 1) random variable.

For alternative #2, the probability that its true value is above 5.5 will be the same, just because of the way we have set up the numbers in this example. We can calculate

$$P\left(V_2 \geq 5.5\right) = P\left(Z \geq \frac{5.5 - 4.5}{1}\right) = P\left(Z \leq -1\right) = 0.1587.$$

However, we can observe from Figure 4.4b that belief density #2 puts more weight on larger values than density #1. For example, the probability that the value of #2 is above 6.5 is larger than the analogous probability for #1. Thus, both #1 and #2 have the same probability of being *better* than #3, but the *magnitude* of this improvement (if it occurs) will tend to be greater for alternative #2. Thus, although we currently think that #1 is better than #2, it is actually #2 that has the highest potential for improvement.

There is also one more way in which new information has the potential to change our beliefs about the alternatives. We could choose to learn about #3 and find that it performs much worse than expected, to the point where #1 begins to look like a better option. The probability that this will occur is

$$P\left(V_3 \leq 5\right) = P\left(Z \leq \frac{5 - 5.5}{0.25}\right) = P\left(Z \leq -2\right) = 0.0228.$$

Note that our *decision* will only change if the value of #3 turns out to be smaller than 5. If our estimate for #3 goes down to 5.25, this will affect our belief about the return on alternative #3, but we will still continue to believe that #3 is better than the others. Likewise, if we learn about alternative #1, our decision will only change if our beliefs about #1 go over 5.5. This in itself is a major insight of optimal learning: new information is only valuable if it changes our decision, and furthermore, the value of information is greater if it is more likely to change our decision.

Each of the three probabilities calculated earlier describes a scenario that changes our beliefs about which alternative is the best. In the first two cases, alternative #1 or #2 performs so well that it begins to outpace #3. In the last case, alternative #3 performs so poorly that #1 becomes the best. In all three cases, the potential of an alternative to change our beliefs about the problem depends on how it compares to the *best of the rest*. An alternative that is not currently our top choice should be evaluated based on how likely it is to improve on that choice. The top choice itself should be compared to the second best. To formalize this concept, we define

$$d_i = \left| v_i - \max_{j \neq i} v_j \right| \tag{4.9}$$

to be the *amount of change* in our beliefs about alternative i needed in order to change our decision. The maximum in (4.9) represents the best of the rest, that is, the highest estimated value among all alternatives other than i. For our example, we can calculate

$$d_1 = |v_1 - v_3| = 0.5$$
$$d_2 = |v_2 - v_3| = 1$$
$$d_3 = |v_3 - v_1| = 0.5$$

as before. Notice that for alternatives #3 and #1 (currently believed to be the best and second best), the needed amount of change is the same: either our estimate for #1 should increase by 0.5 and go above #3, or our estimate for #3 should decrease by 0.5 and go below #1. As the name implies, if the needed amount of change for alternative i is small; learning about this alternative is more likely to change our decision.

However, as we discussed earlier, alternative #2 may have a higher information potential than #1, even though the needed amount of change is greater for #2. To formalize this concept, we introduce a measure of the accuracy of the new information we collect. Define the *uncertainty reduction* for alternative i to be

$$\bar{s}_i^2 = s_i^2 - \left(s_i^{\text{post}} \right)^2, \tag{4.10}$$

where s_i is the standard deviation (uncertainty) of our current beliefs about i and s_i^{post} is the *new* uncertainty about alternative i resulting from the new information. By convention, uncertainty is usually measured using the square of the standard deviation (variance). Then, the difference between these two variances is precisely the amount by which the new information reduces uncertainty.

We have not talked about s_i^{post} thus far, and we cannot calculate it from the numbers given in our example. This quantity is an additional facet of the problem that may have to be estimated by the decision-maker. Consider the following example. For alternative #1, our estimate is equal to 5 with a standard deviation of 0.5. Recall that for normal distributions, a 95% confidence interval for the true value can be constructed by taking

$$m_i \pm 2s_i,$$

so we are 95% confident that the true value is within $1M of the current estimate of $5M. Suppose that the new information (field study, simulation experiment, test market, etc.) can cut this error margin in half: with the new information, the true value will be within $0.5M of the new estimate, whatever that new estimate will be. Then, $s_1^{\text{post}} = 0.25$ and

$$\bar{s}_1^2 = s_1^2 - \left(s_1^{\text{post}}\right)^2 = 0.1875.$$

Similarly, for alternative #2, the standard deviation of our beliefs is $1M, resulting in a margin of error of plus or minus $2M. Suppose that a new experiment on alternative #2 can cut this in half; then, $s_2^{\text{post}} = 0.5$ and

$$\bar{s}_2^2 = s_2^2 - \left(s_2^{\text{post}}\right)^2 = 0.75.$$

In general, if we put the same amount of effort and resources into collecting information, we will achieve more uncertainty reduction about alternatives whose standard deviation is higher. Intuitively, a small amount of field data about a new product or design, for which historical data were unavailable, will have more impact than the same amount of field data about a product with a long history.

If we think about uncertainty reduction in terms of the "margin for error" in our beliefs before and after the new information, we can then use (4.10) to quickly calculate $\bar{s}_i^2$. As a special case, suppose that we are able to learn the *exact* value of alternative i from the experiment (i.e., the new information has perfect accuracy). In this case, we eliminate all of our uncertainty about that alternative, so the variance reduction is simply given by

$$\overline{s}_i^2 = s_i^2. \tag{4.11}$$

In practice, we would use (4.11) in situations where we place a high level of trust in the accuracy of the information we collect and believe that the margin of error is so small as to be negligible.

We now combine (4.9) and (4.10) to define the *normalized change*

$$\overline{d}_i = \frac{d_i}{\overline{s}_i}. \tag{4.12}$$

This quantity compensates a large amount of needed change by the quality of new information. Essentially, if the quality of our experiment is high, it will be easier to discover whether an alternative is significantly better or worse than we thought. For our example, we have

$$\overline{d}_1 = \frac{0.5}{0.1875} = 2.67$$

$$\overline{d}_2 = \frac{1}{0.75} = 1.33.$$

Although the needed amount of change is greater for #2, the *normalized change* is actually smaller, meaning that learning about #2 will be more likely to change our decision.

Finally, we evaluate the normalized change by plugging it into what we will call the *information potential function*. For any real number z, define

$$g(z) = z \cdot F(z) + f(z), \tag{4.13}$$

where F is the CDF of the standard normal (mean 0, variance 1) distribution and f is the standard normal density. Finally, for every alternative i, calculate

$$K_i = \overline{s}_i g\left(-\overline{d}_i\right). \tag{4.14}$$

The alternative with the highest value of information is also the one with the highest value of K_i. This is the alternative that we prefer to learn about.

We will not delve into the mathematical foundation behind (4.13) and (4.14) here, but we provide some intuition. Let us go back to Figure 4.4. Recall that the shaded regions represent scenarios where we learn about one of the alternatives, and the outcome of our experiment changes our beliefs enough to affect our final implementation decision. We can calculate the probability that such a scenario will occur, but this number does not capture certain nuances: recall that, although the probabilities were the same for alternatives #1 and #2, the belief density for #2 placed more weight on larger values. Thus, if #2 turned out to be the best alternative, the exact value of the "improvement" thus obtained would also tend to be larger than if #1 turned out to be the best. The information potential function g is a formula derived by computing the weighted average of the improvement. Essentially (omitting some details), we are adding up all the values in the shaded region and weighting each one by the likelihood that it will be the true value. The formula computed in (4.14) has several intuitive properties following from this concept:

- If d_i is smaller, K_i will be larger. Recall that, if the amount of needed change is small, alternative i has more potential to change our decision. A result of this is that if the estimated value of alternative i is far below the current best alternative, the needed amount of change will be much greater, and we will be less likely to consider it.
- If $\bar{s}_i$ is larger, K_i will be larger. Uncertainty reduction is inherently valuable because it leads to more accurate beliefs and decisions. An alternative that seems to be poor but has a large amount of uncertainty is more likely to be much better than we think.
- If $\bar{d}_i$ is smaller, K_i will be larger. Again, if the normalized change is small, alternative i has more potential to change our decision.

All together, (4.14) provides a balance between our current beliefs about an alternative and our uncertainty about that alternative. The formula (known by the names "expected improvement" and "knowledge gradient" and originally developed by Gupta & Miescke, 1996) also considers the relationship between alternative i and the other available decisions.

Although the mathematical reasoning behind (4.14) is somewhat involved, the calculations needed to make decisions are much simpler and can easily be carried out in a spreadsheet. For our running example with three alternatives, the calculations are given in the following table.

Alternative	m_i	s_i	d_i	$\bar{s}_i$	$\bar{d}_i$	K_i
1	5	0.5	0.5	0.4330	1.1547	0.0266
2	4.5	1	1	0.8660	1.1547	0.0532
3	5.5	0.25	0.5	0.2165	0.0036	0.0008

The learning decision is made by checking the final column. Alternative #2 is found to have the highest value of information, despite having the lowest estimated beliefs, due to a high degree of uncertainty and uncertainty reduction. Thus, we choose to learn about #2 before making our final implementation decision. Note that alternative #3, despite having the highest estimated beliefs, has the lowest information potential. This simply means that we prefer to learn about #2; however, if the results of our experiment are insufficiently strong, we may still prefer to implement #3.

One major question remains. So far, we have discussed how to efficiently use a single opportunity to collect information. We have assumed that we can choose one alternative to learn about, but that the final implementation decision will be made after this new information is collected. However, it is more likely that we will have more than one chance to learn before having to commit to an implementation. For example, our budget may allow for more than one test market. If we are using simulation models to evaluate the performance of our alternatives, we would typically set aside several weeks or months of simulation time, in order to learn as much about as many alternatives as possible. In short, the information we collect does not have to immediately lead into the implementation decision. Rather, it should feed back into our beliefs and prepare us for the next opportunity to collect information. Thus, we need a simple mechanism for *updating* our beliefs with new information, similar to the one used in the newsvendor problem to update a gamma distribution.

The updating mechanism is easy to use, but requires us to specify an additional input s_i^{noise} representing the accuracy of the new information collected, in the form of a standard error. We call this the *noise error*, and it simply means that, on average, the new information will be accurate (if we conduct many hundreds or thousands of experiments and average the output, we will learn the correct value of the alternative), but a single experiment may have some variation. We are 95% confident that the output of the experiment will be equal to the true value, plus or minus some "noise" whose magnitude is within $2s_i^{\text{noise}}$. Once we specify s_i^{noise}, we no longer have to come up with a value for s_i^{post}, the new uncertainty of our beliefs after the experiment. This quantity is given by a formula

$$\left(s_i^{\text{post}}\right)^2 = \frac{1}{\dfrac{1}{s_i^2} + \dfrac{1}{\left(s_i^{\text{noise}}\right)^2}}. \tag{4.15}$$

Although (4.15) seems complicated, its meaning is actually quite simple. Since standard deviation (or its square, the variance) represents uncertainty, it follows that larger s means greater uncertainty. Then, the reciprocal

$$p_i = \frac{1}{s_i^2}$$

can be viewed as a measure of the *precision* of our beliefs—larger s means smaller p and thus less precision. Then, (4.15) can be rewritten as

$$p_i^{\text{post}} = p_i + p_i^{\text{noise}},$$

which implies that our beliefs always become more precise when we obtain new information (alternately, "more information is always useful"). With regard to our actual estimate of the value, it changes according to the equation

$$v_i^{\text{post}} = \frac{p_i v_i + p_i^{\text{noise}} X_i}{p_i + p_i^{\text{noise}}}, \tag{4.16}$$

where X_i is the observed output of the experiment (e.g., the sales in the test market or the simulated performance of the design). Equation (4.16) is simply a weighted average of our old beliefs about alternative i, represented by v_i, and the new information about alternative i that we have obtained from the experiment, represented by X_i. The weight of the old beliefs is determined by the precision of those beliefs, represented by p_i. Thus, if our original beliefs were more precise (our original uncertainty about the value of i was smaller), we will rely on them more, and our new beliefs v_i^{post} will be closer to their original value v_i. If our original beliefs were imprecise (high uncertainty), we will rely more on the new information obtained from experiments.

Finally, we can put everything together into a procedure for collecting information and making decisions. The procedure will go as follows:

- Begin with a set of beliefs, consisting of two numbers v_i and s_i, about every alternative i.
- Based on the current beliefs, calculate the information potential K_i from (4.14), for every i.
- Perform an experiment on the alternative with the highest information potential. If this is alternative i, collect a new observation or performance value X_i from the experiment.
- For the alternative we measured, update the beliefs using Equations (4.15) and (4.16). For the alternatives that were not measured, leave the beliefs as they were.
- Repeat as many times as necessary, while our budget allows us to perform experiments. For each experiment, calculate K_i using only the most recent beliefs.
- After all experiments have been used, implement the alternative with the highest estimated value, based on the most recent beliefs.

At this time, there is a considerable volume of simulated experiments (Frazier, Powell, and Dayanik, 2008; Negoescu, Frazier, and Powell, 2011; Ryzhov, Powell, and Frazier, 2012) suggesting that different versions of this procedure offer significant advantages over other strategies and rules of thumb such as interval estimation. The main advantage is that the concept of information potential provides a way to quantify the exact balance between "learning" (the value of new information) and "earning" (the need to find a good solution). By looking at the way this quantity is calculated, we can obtain several important insights about the problem of learning:

- The value of learning about an alternative is directly related to how likely that alternative is to change our decision. New data are only relevant if they have the ability to change the way decisions are made.
- An alternative is more likely to be important (i.e., have high potential to change our decision) if it is believed to high value, perhaps one that is close to the current best candidate.

- However, an alternative is also more likely to be important if we have more uncertainty about it, because this increases the odds that its actual value is much better than our old data suggest. The estimated value (obtained from the old data) is balanced with uncertainty in the "normalized change" calculation.

One important point to keep in mind is that it is crucial for the decision-maker to have a realistic sense of how much uncertainty is present in the problem. If we are overconfident about an alternative and set s_i to be much smaller than it should be, this will affect the performance of the procedure. If we do not have a clear sense of the margins of error for our current beliefs, it is better to err on the side of caution and assume more uncertainty in the problem.

OPTIMIZING A RULE-BASED POLICY FOR INVENTORY MANAGEMENT

We now give one example of how the selection procedure can be used in practice to optimize a decision. Our discussion in the preceding section has used the abstract term "alternative" to describe any choice available to the decision-maker. This term can be defined quite broadly. In fact, we can integrate our learning procedure into various "rule-based" strategies for making decisions. Instead of forcing a decision-maker to adopt a certain practice, we can leverage optimal learning to optimize or improve an existing practice. We illustrate this with an example from inventory management, namely, the (q, Q) inventory model of Scarf (1960).[2]

Suppose that a single product is stored at a warehouse. Initially, S_0 units are in stock. During the week, demand for the product is observed and the inventory decreases. The total demand during the first week is denoted by Z_1. All the demand must be satisfied; if the warehouse runs out of inventory before the week is over, it is necessary to back-order additional units (at an extra cost to the company). The warehouse manager also has the option to place an order for additional inventory at the beginning of the week (for ease of explanation, let us assume

---

[2] This model is also commonly referred to as the (s, S) model, but the meaning remains the same.

that it arrives instantaneously), but the manager has no way of knowing what the demand for the coming week will be. There is a fixed and variable cost for ordering, and unsold inventory also incurs a holding cost if carried over to the next week, so the manager would prefer not to order as long as there is currently enough to avoid a stock-out. At the same time, waiting too long to order increases the risk of these expensive stock-outs.

Clearly, the manager needs a strategy for determining when orders should be made and how much should be ordered. The insight of Scarf (1960) is that the total cost (holding, ordering, and shortage) can be minimized by a simple strategy described in Figure 4.5. First, we choose two numbers q and Q satisfying $0 < q < Q$. We then apply the following rule:

- If, at the beginning of a new week, the current inventory level is below q, order exactly enough to bring it up to Q. That is, if there are 3 boxes in inventory, q is 5 boxes, and Q is 10 boxes, the manager should order 7 boxes.
- Otherwise, do not order anything and wait 1 week.

The blue line in Figure 4.5 is the changing inventory level. We see that the initial inventory level S_0, at the beginning of week 0, is between q and Q, so the manager does nothing. By the beginning of week 1, the inventory level S_1 has dropped below q, so the manager brings it up to Q. During the next two weeks, the inventory decreases but remains above q, so no further action is required. Essentially, we only reorder

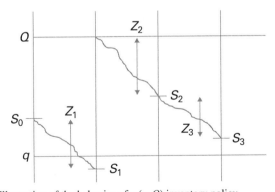

Figure 4.5   Illustration of the behavior of a (q, Q) inventory policy.

inventory when it becomes "low enough," and then we order just enough to make it "high enough."

In one of the most celebrated results of operations management, Scarf (1960) proved that this simple strategy can be *optimal*, that is, it can minimize total long-term cost. The catch, however, is that the two numbers q and Q have to be chosen correctly. For the sake of illustration, suppose that the fixed ordering cost is $0.15 per order (regardless of how many boxes are ordered), the holding cost is $0.35 to store one box for one additional week, the shortage cost is $1.20 for each back-ordered box, and the order cost is $0.90 per box. Unfortunately, Scarf's proof does not provide us with a way to find the "best" settings of q and Q. In order to truly minimize the total cost, these numbers must be carefully tuned.

The only way to learn about the performance of a particular choice of q and Q is to experiment with them. Either we could implement a (q, Q) strategy with our chosen two values in the field and observe the results after several weeks, or we could run a simulation model with the same values, using historical demand data to model the random weekly fluctuations in the demand. Trial and error is less costly if we do it inside a simulator rather than directly in the field, but nonetheless, we may not have enough time or computational resources to try every conceivable value of q and Q. Keep in mind also that other, more complex management problems may involve strategies with more than two tunable "parameters" (e.g., we may wish to assign high-priority calls to call center workers whose performance scores are above a certain threshold) making it even more difficult to find the right configuration.

Suppose that we are considering three possible values for each parameter (low, medium, and high). Thus, there are six possible combinations of q and Q (since Q must always be greater than q). The following table shows the performance of each configuration, in terms of total cost over some fixed length of time. These performance values were found by running a large number of simulations, with the assumption that weekly demands are independent and follow a normal distribution with a mean of 5 and standard deviation of 0.25. In practice, we would not know these demands and would have to infer the distribution from data—but even if we were to know the distribution *exactly*, there is still no easy way to calculate the optimal q and Q.

Alternative	q	Q	Total cost (1 simulation)	Total cost (1000 simulations)
1	1	2	$3852	$3656
2	1	5	$2895	$2473
3	4	5	$2792	$2498
4	1	10	$2751	$3053
5	5	10	$3636	$3541
6	9	10	$4156	$4264

From the last column of the table, we can see that the optimal choice seems to be configuration #2, where $q=1, Q=5$, narrowly beating out $q=4, Q=5$. However, these numbers were obtained by running 1000 simulations for each configuration. If the total number of configurations were larger (e.g., several hundred or thousand), this exhaustive study would not be feasible. We would have to do the best we could with a limited number of simulations. The challenge is that a small number of simulations may produce inaccurate results: the above table shows that when we only perform one simulation on each configuration, the true best configuration #2 actually appears to be the third best, whereas the true third-best configuration #4 seems to be the best!

This is precisely the type of setting where optimal learning can help. Using the language of the previous section, let v_i, our initial beliefs about configuration i, be equal to the number obtained from a *single* simulation, as given in the above table (so, $v_1=3852, v_2=2895$, etc.). Let s_i be equal to $1000, suggesting that we believe the *true* values to be within plus or minus $2000 of the initial estimates (which certainly seems reasonable). Let s_i^{noise} be equal to $500, suggesting that the outcome of any individual simulation will be between plus or minus $1000 of the true value (a conservative estimate, considering the numbers in the table). Note that these are very rough guidelines for the uncertainty, but as we shall see, they are quite sufficient. While we would get better results by crafting these numbers more carefully, optimal learning is often able to discover optimal solutions starting from very rough beliefs.

Now, we can calculate the information potential. For the initial values as described earlier, we would make the calculations given in the following table. Note that since we are *minimizing* cost rather than *maximizing* value, we repeat the analysis of the preceding section under the assumption that smaller values are better. Thus, (4.9) becomes

$$d_i = \left| v_i - \min_{j \neq i} v_j \right|,$$

as the "best of the rest" is now the one with the smallest value.

Alternative	m_i	s_i	d_i	$\bar{s}_i$	$\bar{d}_i$	K_i
1	3852	1000	1101	970.1425	1.1349	62.1062
2	2895	1000	144	970.1425	0.1484	319.2866
3	2792	1000	41	970.1425	0.0423	366.8764
4	2751	1000	41	970.1425	0.0423	366.8764
5	3636	1000	885	970.1425	0.9122	95.2667
6	4156	1000	1405	970.1425	1.4482	31.9590

Because s_i and s_i^{noise} are the same for every alternative, the uncertainty reduction $\bar{s}_i$ is also the same (and also fairly high, considering the large difference between the belief and observation uncertainties). We do not yet have enough information to discern that we should look into configuration #2. However, even with our inaccurate initial beliefs, we see that the apparent best and second-best configurations (#3 and #4) have the same information potential. Configuration #2 is not far behind; the gap in potential is smaller than the difference in the initial estimates might indicate. Still, the choice of how to use the first simulation appears to be tied between #3 and #4, when we saw before that #2 is the best.

However, recall that the true power of optimal learning is in its ability to adapt its recommendations to new information. Having decided to learn about #3 or #4, we can run a single simulation of that configuration, then use (4.15) and (4.16) to update our beliefs with new information, and recalculate K_i with the new beliefs. Suppose that our total information budget consists of 10 simulations. The following table shows our *final* beliefs about each configuration, after allocating 10 simulations to configurations with high information potential. Since the outcome of each simulation is random (due to the randomly generated demands), there will also be some variation in our final beliefs. To get a sense of the sensitivity of our solution to the random fluctuations in the demands, we run this problem five different times, starting from the same initial beliefs every time. The numbers in parentheses represent the number of simulations that were allocated to a configuration in a particular run.

Alternative	Run 1	Run 2	Run 3	Run 4	Run 5
1	3585 (1)	3852 (0)	3852 (0)	3520 (1)	3787 (1)
2	*2519 (3)*	*2461 (4)*	*2402 (4)*	*2598 (3)*	*2510 (4)*
3	2526 (4)	2507 (4)	2441 (4)	2686 (3)	2629 (3)
4	2942 (1)	2949 (1)	2981 (1)	2967 (2)	3000 (1)
5	4026 (1)	4137 (1)	3038 (1)	3153 (1)	3708 (1)
6	4156 (0)	4156 (0)	4156 (0)	4156 (0)	4156 (0)

Notice that although none of the final beliefs are exactly equal to the true values of the configurations (the values computed using 1000 simulations), they are enough to correctly identify #2 as the best configuration in *all five* runs. Note also that in all five runs, the majority of the information budget (60–80%) is divided between alternatives #2 and #3, the true best and second best. Occasionally, we use a simulation to learn about #1, #4, or #5, but stop learning about them once we see that they are not promising. Configuration #6 simply does not show enough potential to ever justify the cost of learning about it.

By contrast, suppose that we follow the strategy of always simulating whichever configuration appears to be the best (has the lowest estimated value) at any given moment, with no regard for uncertainty or information potential. The final beliefs obtained using this "greedy" or "myopic" strategy are shown in the following table.

Alternative	Run 1	Run 2	Run 3	Run 4	Run 5
1	3852 (0)	3852 (0)	3852 (0)	3852 (0)	3852 (0)
2	2895 (0)	2895 (0)	2895 (0)	2895 (0)	2895 (0)
3	*2454 (9)*	*2444 (9)*	*2467 (8)*	*2479 (9)*	*2600 (9)*
4	2937 (1)	2843 (1)	2830 (2)	2973 (1)	3095 (1)
5	3636 (0)	3636 (0)	3636 (0)	3636 (0)	3636 (0)
6	4156 (0)	4156 (0)	4156 (0)	4156 (0)	4156 (0)

The situation is now quite different. We begin by measuring #4, since it appears to be the best according to our initial beliefs. Then, we quickly learn that it is not as good as it seems and switch to #3, which we initially believed to be the second best. As we learn about #3, we discover that it is indeed reasonably good, so we stick with it and never choose to learn about #2. Our final decision after 10 simulations would be to implement #3 in all five runs, although in reality this is the second-best configuration. By considering the information potential of the different configurations, we were able to engage in more experimentation and learn something about 4–5 of the available alternatives, rather than only 1–2. However, it was not simply arbitrary trial and error, but rather targeted experimentation: we spent most of our learning budget on the top two candidates while occasionally learning something about several others.

What if we were to learn by simply simulating randomly chosen configurations? This is equivalent to splitting our information budget equally among the alternatives (on average) and would seem like a reasonable approach if we really had no idea which configuration was the best. Our last table gives the results for this approach.

Alternative	Run 1	Run 2	Run 3	Run 4	Run 5
1	3717 (3)	3763 (2)	3807 (1)	3852 (0)	3527 (2)
2	2895 (0)	*2433 (4)*	*2366 (2)*	2586 (1)	2895 (0)
3	*2459 (2)*	2792 (0)	2529 (1)	*2506 (1)*	*2483 (3)*
4	2927 (1)	2810 (1)	3058 (2)	3073 (4)	3027 (2)
5	2956 (1)	3076 (1)	3328 (2)	3641 (2)	4023 (1)
6	4207 (3)	4150 (2)	4391 (2)	4104 (2)	4273 (2)

This seems to do a little better—we now find the best configuration in two out of five runs. The other three times, we still pick #3. The drawback of this method is that we are wasting a significant portion of our budget on alternatives that don't matter, such as #6. Earlier, we saw that there was simply no reason to learn about this configuration, as it was just too poor to ever have a chance of being the best. Also, if the number of alternatives is large, spreading out the budget will be quite ineffective; consider, for example, the task of allocating 10 simulations among 1000 alternatives.

The main message of optimal learning is that it is possible to optimize the collection of information, just as a business would optimize the use of any other resource. When faced with the problem of learning about a number of unknowns, we should strike a balance between learning about solutions that seem to be good and learning about solutions that seem to be suboptimal yet have high potential to be better than we think. Although the aforementioned example was small enough to allow us to find the best configuration by brute force (running 1000 simulations on every alternative), it is easy to envision problems where the number of alternatives is in the hundreds or thousands. For example, if we are designing a new marketing campaign with 10 possible design attributes, where we have the choice of including or not including each attribute in the campaign, the number of possible combinations of attributes is equal to $2^{10} = 1024$. Furthermore, in that setting, the way to collect information would be, not to run a simulation model, but rather to conduct a limited field test with a particular design, avoiding the need to do thousands of experiments.

DISCUSSION

As you read this chapter, please keep in mind that, like any concept from operations management, optimal learning is not limited to the two specific models we have covered here. Many of the applications we have cited require mathematical tools going beyond what we showed here. We have used these two models as examples to illustrate several key concepts that arise in every optimal learning problem and are addressed in some way by any learning model. These include:

- *Every model is uncertain.* Tools such as decision analysis help us to consider multiple uncertain outcomes (such as whether the demand will be high or low) during planning, but even these tools are dependent on some model for how likely one outcome is over another. These models are never completely accurate; rather, they are improved over time as new data come in.
- *Model uncertainty changes decisions.* An ordering decision may be based on some forecast of demand. If new data comes in to alter the forecast, the decision will change as well. The best decisions are those that can anticipate the possibility of such change, before it occurs.
- *Information is only valuable if it changes decisions.* Collecting data about an alternative that is clearly suboptimal is unproductive. Rather, it is better to spend a limited information budget on alternatives that

have high potential. This does not necessarily mean that they look good to us now, only that there is a reasonable chance that they might be better than we think.

- *Uncertainty is valuable.* It is more difficult to make good decisions in an uncertain world, but even uncertainty can be an asset. If we are uncertain about a decision, that in itself gives it higher potential to be good. If we can make an informed trade-off between this potential and our need to optimize quickly, we increase our ability to discover a game-changing new decision.

Optimal learning methods work best when the information budget is very limited (if we have the ability to perform thousands of experiments, it is less important to worry about allocating them efficiently), but the number of alternatives is reasonably large (perhaps several thousand). As is frequent in operations research, optimal learning is most effective when used in tandem with other methodologies such as data analytics to build a body of prior information and conduct a prescreening of obviously unpromising choices. It generally works well with simple models for representing beliefs (e.g., using means and standard deviations), to minimize the time required to calculate information potential and other relevant quantities. However, optimal learning can leverage a simple belief model to develop sophisticated metrics for valuating alternatives and information that clearly identify the most promising alternatives and information, providing decision-makers with guidance in uncertain environments.

REFERENCES

Arlotto, A., Chick, S.E., and Gans, N. (2010) "Optimal employee retention when inferring unknown learning curves." In: *Proceedings of the 2010 Winter Simulation Conference* (eds. B. Johansson, S. Jain, J. Montoya-Torres, J. Hugan, E. Yücesan), Institute of Electrical and Electronics Engineers, Inc., Piscataway, NJ, pp. 1178–1188.

Eneyo, E.S. and Pannirselvam, G.P. (1998) "The use of simulation in facility layout design: a practical consulting experience." In: *Proceedings of the 1998 Winter Simulation Conference* (eds. D.J. Medeiros, E.F. Watson, J.F. Carson, M.S. Manivannan), Institute of Electrical and Electronics Engineers, Inc., Piscataway, NJ, pp. 1527–1532.

Frazier, P.I., Powell, W.B., and Dayanik, S. (2008) "A knowledge-gradient policy for sequential information collection." *SIAM Journal on Control and Optimization* **47**(5), 2410–2439.

Gupta, S. and Miescke, K. (1996) "Bayesian look ahead one-stage sampling allocations for selection of the best population." *Journal of Statistical Planning and Inference* **54**(2), 229–244.

Kaelbling, L.P. (1993) *Learning in Embedded Systems*. MIT Press, Cambridge, MA.

Lariviere, M.A. and Porteus, E.L. (1999) "Stalking information: Bayesian inventory management with unobserved lost sales." *Management Science* **45**(3), 346–363.

Marmidis, G., Lazarou, S., and Pyrgioti, E. (2008) "Optimal placement of turbines in a wind park using Monte Carlo simulation." *Renewable Energy* **33**(7), 1455–1460.

Negoescu, D.M., Frazier, P.I., and Powell, W.B. (2011) "The knowledge-gradient algorithm for sequencing experiments in drug discovery." *INFORMS Journal on Computing* **23**(3), 346–363.

Powell, W.B. and Ryzhov, I.O. (2012) *Optimal learning*. John Wiley & Sons, Inc., Hoboken, NJ.

Ryzhov, I.O., Powell, W.B., and Frazier, P.I. (2012) "The knowledge gradient algorithm for a general class of online learning problems." *Operations Research* **60**(1), 180–195.

Sashihara, S. (2011) *The Optimization Edge: Reinventing Decision Making to Maximize All Your Company's Assets*. McGraw-Hill, New York.

Scarf, H.E. (1960) "The optimality of (s, S) policies in the dynamic inventory problem." In: *Mathematical Methods in Social Sciences* (eds. J. Arrow, S. Karlin, P. Suppes), Stanford University Press, Stanford, CA , pp. 196–202.

Stevenson, W.J. (2008) *Introduction to Operations Management* (10th edn.). McGraw-Hill, Singapore.

Wang, Y., Gilland, W., and Tomlin, B. (2010) "Mitigating supply risk: dual sourcing or process improvement?" *Manufacturing & Service Operations Management* **12**(3), 489–510.

CHAPTER 5

Using Preference Orderings to Make Quantitative Trade-Offs

Chen Wang[1] **and Vicki M. Bier**[2]
[1] *Department of Industrial Engineering, Tsinghua University, Beijing, P.R., China*
[2] *Department of Industrial and Systems Engineering, University of Wisconsin-Madison, Madison, WI, USA*

INTRODUCTION

Real-world decision-makers often face multiple objectives. For example, many decisions in the business context aim to maximize both profits and market share (Keeney, 2007), while consumers may want to trade off the quality of a product against its price. Multiattribute utility theory (MAUT; Keeney and Raiffa, 1976) is one of the most well-known and rigorous methods for trading off multiple (generally conflicting) objectives. The additive form of the multiattribute utility function is often used, due to its simplicity, and has been shown to give satisfactory results in a wide variety of circumstances (Dawes and Corrigan, 1974; Stewart, 1996).

Breakthroughs in Decision Science and Risk Analysis, First Edition.
Edited by Louis Anthony Cox, Jr.

One important task in constructing multiattribute utility functions is to choose appropriate attributes to quantitatively reflect stakeholder objectives. Desirable attributes should be measurable on either natural scales (e.g., lives saved, dollars spent) or constructed scales. For an example of a constructed scale, Keeney and von Winterfeldt (1994) use a 0–5 scale to measure the environmental impact of nuclear waste disposal strategies. Attributes should also directly and unambiguously describe the degree to which the decision-maker's objectives are met (Clemen and Reilly, 2001).

Another key task in the development of a multiattribute utility function is to obtain estimates for the weights (i.e., the relative importance) of the various attributes. Traditional elicitation methods include the ratio method (Edwards, 1977), the swing-weight method (von Winterfeldt and Edwards, 1986), and the trade-off and pricing-out methods (Keeney and Raiffa, 1976). However, these methods can be quite time-consuming to use. Moreover, asking stakeholders to provide precise assessments of attribute weights may also imply unwarranted precision of the elicitation results, due to the inherent uncertainties associated with subjective judgments (Schoemaker and Waid, 1982; Borcherding, Eppel, and von Winterfeldt, 1991).

Providing ordinal information rather than cardinal assessments is widely believed to be easier and more reliable. In fact, making pairwise comparisons has been suggested to be the first step in the process of evaluating alternatives (Watson and Buede, 1987). Moreover, ordinal judgments are generally easy to understand and interpret, avoiding the need for extensive orientation of stakeholders prior to the elicitation process. Finally, while it is unrealistic to expect agreement on cardinal assessments in a group decision process, consensus on rankings is often attainable (Kirkwood and Sarin, 1985). There have also been a broad range of applications on eliciting multiattribute preferences from ordinal information, for example, in the fields of marketing research (Green and Srinivasan, 1978; Green, Krieger, and Wind, 2001), health economics (McCabe et al., 2006; Ali and Ronaldson, 2012), and adversary preference modeling (CREATE, 2011).

Typical ordinal data for constructing multiattribute utility functions include rank orderings (or pairwise comparisons) of attribute weights, alternatives, or utility differences between paired alternatives (Horsky and Rao, 1984). For example, the Simple Multi-Attribute Rating Technique Extended to Ranking (SMARTER) method (Edwards and Barron, 1994) uses rank orderings of attribute importance directly to derive point estimates for the attribute weights. By contrast, the maximum entropy utility method (Abbas, 2006) uses rank orderings of the attractiveness of alternatives to construct an overall value function for the alternatives. This method also has the potential for handling ordinal judgments over utility differences between paired alternatives; however, it does not provide estimates for the attribute weights.

In this chapter, we focus on an indirect elicitation process that converts (partial) rank orderings of the alternatives into point estimates or probability distributions for the various attribute weights. In particular, we discuss three mathematical methods that can be used to derive weight estimates in the additive multiattribute utility from rank orderings of alternatives: conjoint analysis in marketing (Shocker and Srinivasan, 1973; Green and Srinivasan, 1978), probabilistic inversion (PI; Bedford and Cooke, 2001; Neslo et al., 2011), and Bayesian density estimation (BDE; Ferguson, 1973). Conjoint analysis is compelling because of its simplicity and computational ease, at least in its most widely used form of the Linear Programming Technique for Multidimensional Analysis of Preference (LINMAP) (Green and Srinivasan, 1978). However, LINMAP generates only point estimates for the attribute weights, not probability distributions. Some conjoint approaches apply mixed-logit models to produce distributions for the various weights to reflect disagreements and variations across different stakeholders, but these often assume that the attribute weights are normally distributed (McFadden and Train, 2000).

The ability to generate probability distributions (rather than only point estimates) is if anything even more important for an elicitation method that deals with ordinal data than for one that deals with cardinal data. In general, ordinal judgments are relatively weak, so it is important for an elicitation method to reflect the nature and degree of uncertainties arising from the use of ordinal data. Moreover, since single-modal parametric distributions are generally not adequately flexible to handle widely differing, preferences, mixtures of distributions or other nonparametric representations may be preferable in practice. Therefore, PI and BDE have recently been adapted to apply to ordinal preference rankings (Neslo et al., 2011; Wang and Bier, 2013). Both PI and BDE can explicitly capture stakeholder uncertainties and disagreements, by generating probability distributions over attribute weights without parametric assumptions such as normality.

Additionally, if a stakeholder's rank orderings of alternatives cannot be explained by the available set of attributes, then the LINMAP approach of conjoint analysis might give weight estimates that do not even approximately reproduce the stated rank orderings of the alternatives. Other conjoint approaches based on logit or mixed-logit regression use a large variance of the residuals to reflect the lack of consistency between stakeholder preferences and the given set of attributes, but these approaches do not provide a unit-free measure to quantify the degree of inconsistency. By contrast, inspired by Jenelius, Westin, and Holmgren (2010), PI and BDE as applied in Wang and Bier (2013) allow for an unobserved attribute that may be important to the stakeholders but has not been included in the analysis. The weight assigned to this unobserved attribute thus indicates how well or poorly the available set of attributes fits the given stakeholders' judgments. Moreover, estimated values of the unobserved attribute for the various alternatives can also help to indicate which attributes may be missing.

In principle, conjoint analysis, PI, and BDE require stakeholders to provide only rank orderings of alternatives, which are believed to be easy to assess and interpret. Therefore, these methods may be quite useful in eliciting preferences of stakeholders who do not have highly quantitative backgrounds. Moreover, the simplicity and automatic consideration of stakeholder uncertainties and disagreements would make these methods especially powerful in large-scale elicitation tasks (e.g., using online surveys).

Note that direct elicitation may still be necessary for making high-stakes decisions such as major public policies, since stakeholders are more likely to act on recommendations based on the elicitation results if they feel that their views have been treated seriously and fully incorporated (Keeney, 2007). However, conjoint analysis, PI, and BDE can all be used not only as alternatives to direct elicitation but also as a source of inputs for convergent validation, a common technique in decision analysis. For example, we could ask the stakeholders to comment on any observed discrepancies between the results of different elicitation approaches—for example, whether they put more credence in attribute weights inferred from their rankings of alternatives or in their directly assessed attribute weights.

In the following, we first provide a literature review on existing ordinal elicitation methods (including both direct and indirect approaches) and give some background on conjoint analysis, PI, and BDE. We then discuss how the three methods estimate attribute weights from rank orderings of alternatives provided by stakeholders, followed by an illustrative case study on adversary preference elicitation. Next, we discuss how to use negative weights to capture the possibility that stakeholders may disagree on the direction of effects of each attribute (following CREATE, 2011), that is, whether a larger value of an attribute corresponds to higher or lower utility.

We then conduct simulation-based sensitivity analysis to show how results differ when using complete rank orderings versus partial rank orderings of the alternatives. This is important, since if stakeholders needed to rank dozens of alternatives in order to obtain adequate weight estimates, ordinal elicitation methods would not be terribly practical. Finally, we conclude the chapter and present some directions for future research.

LITERATURE REVIEW

For simplicity, we focus our attention on the two-attribute case. (The discussion can be extended to cases with more than two attributes in a straightforward manner.) The standard additive multiattribute utility function takes the form

$$U_n = w_1 v_{n1} + w_2 v_{n2} \tag{5.1}$$

Where U_n is the overall utility of alternative n, and v_{nm} is the single-attribute utility of alternative n on attribute m (for $n = 1, \dots, N$ and $m = 1, 2$). The weights of the various attributes are assumed to be non-negative and sum to one; that is, $w_m \geq 0$ for $m = 1, 2$ and $w_1 + w_2 = 1$.

Note that v_{nm} reflects the stakeholder utility of alternative n on attribute m, with 1 representing the best possible value and 0 the worst possible value. Nonlinear single-attribute utility functions can in principle be used to capture stakeholder risk attitudes toward uncertain consequences (e.g., risk aversion, risk neutrality, or risk proneness; see von Winterfeldt and Edwards, 1986). However, detailed elicitation of stakeholder risk attitudes can be quite time-consuming. Instead, we could simply assume that v_{nm} is linear in the original attribute value (the approach adopted in this chapter) or use a logarithmic relationship based on Fechner's law (Fechner, 1860), which states that human perceptions are logarithmic in the magnitude of the stimuli.

Following Wang and Bier (2013), we extend Equation 5.1 to incorporate an unobserved attribute that may be important to the stakeholders, but has not been identified in the elicitation process. Specifically, we let w_3 be the weight for the unobserved attribute and $(y_1, \ldots, y_N)$ be the utilities of the N alternatives on the observed attribute. The revised multiattribute utility function is then given by

$$U_n = w_1 v_{n1} + w_2 v_{n2} + w_3 y_n, \qquad (5.2)$$

where $w_m \geq 0$ for $m = 1, 2, 3$ and $w_1 + w_2 + w_3 = 1$. Again, the single-attribute utilities of the unobserved attribute for the various alternatives y_n are assumed to take on values in [0, 1]. Note that the weight on the unobserved attribute, w_3, can be used to assess whether the two known attributes are adequate to capture the stakeholders' stated (partial) rank orderings of the N alternatives.

A variety of elicitation methods in the literature are based on ordinal data. Some methods provide estimates for the overall utility values U_n directly. For example, Abbas (2006) defines "utility densities" analogous to probability densities and applies the principle of maximum entropy to assign utilities to the alternatives directly according to their ranks. However, this method does not provide explicit estimates of attribute weights and therefore cannot be applied to additional alternatives that have not been ranked.

Other elicitation methods have been developed to generate estimates for the attribute weights w_m. For example, to avoid the difficulties associated with exact weight assessment, some researchers recommend instead simply rank ordering the importance of the various attributes (Dawes and Corrigan, 1974; Stillwell, Seaver, and Edwards, 1981). Among the existing formulas that convert rank orderings of attribute importance into cardinal estimates of attribute weights, SMARTER (Edwards and Barron, 1994) has been shown to be the most accurate (Barron and Barrett, 1996; Sarabando and Dias, 2006). (Note, however, that SMARTER has not been applied to cases with unobserved attributes, perhaps because it seems unrealistic to rank an attribute that we do not even know.) Suppose that a stakeholder weights attribute 1 more heavily than attribute 2; then the resulting weight estimates are the coordinates of the centroid of the region $\{(w_1, w_2) | w_1 + w_2 = 1, w_1 \geq w_2 \geq 0\}$.

Note that SMARTER generates only point estimates of attribute weights rather than probability distributions. Therefore, it may not be appropriate for situations in which uncertainty about preferences is a crucial consideration. However, Rao and Sobel (1980) have derived a marginal distribution for the kth largest weight using the principle of maximum entropy (such that no more information is reflected by the distribution than that given by the ranks only), in which the marginal means of the attribute weights correspond exactly to the SMARTER weights (Barron and Barrett, 1996).

In this chapter, however, we are interested in methods that yield estimates for attribute weights starting from ordinal judgments of alternative attractiveness U_n (rather than ordinal judgments of attribute importance w_m). Conjoint analysis in marketing is one method for doing this (Green and Srinivasan, 1978). A representative approach to conjoint analysis, LINMAP (Shocker and Srinivasan, 1973), uses a linear program to find weight estimates that yield the smallest sum of pairwise "violations" from a single stakeholder's ranking of the alternatives. However, like SMARTER, LINMAP provides only point estimates for the attribute weights. In addition, the optimal attribute weights may not be unique, and there is no agreement in the literature on how to handle multiple optima. Finally, it is an open question how to aggregate attribute weights obtained from multiple stakeholders using this approach.

Another version of conjoint analysis uses logistic regression to obtain weight estimates that best fit a stakeholder's pairwise comparisons of alternatives (McFadden, 1977). However, this approach is based on the unrealistic assumption that different pairwise comparisons of the alternatives (derived from a single rank ordering of all alternatives given by a stakeholder) are independent. Moreover, it also yields only point estimates of attribute weights, which may not be adequate to capture heterogeneity among stakeholders. Mixed-logit models can be applied to obtain probability distributions for the attribute weights but often require that the distribution of attribute weights be multivariate normal (Revelt and Train, 1998; McFadden and Train, 2000).

By contrast, PI and BDE can be used to infer a joint probability distribution over the attribute weights (without distributional assumptions such as normality) from rank orderings of alternatives. Moreover, both PI and BDE take care of the aggregation of heterogeneous stakeholder opinions automatically and explicitly capture stakeholder disagreements by (possibly multimodal) distributions.

In particular, PI (Cooke, 1994; Bedford and Cooke, 2001; Kraan and Bedford, 2005) aims to find a probability distribution over the attribute weights that can reproduce the stakeholders' stated (marginal) rank orderings of alternatives. Neslo et al. (2011) were the first to apply PI to obtain distributions over multiple attribute weights that best fit the ordinal judgments given by stakeholders. Recently, Wang and Bier (2013) extend that model to include an unobserved attribute that may be important but has not yet been identified, which ensures the existence of feasible solutions in PI at least in theory.

The idea of BDE (Ferguson, 1973) is to update a decision-maker's (possibly noninformative) prior distribution over the attribute weights by treating each stakeholder's rank orderings as independent observations. BDE also allows the decision-maker to specify a level of reliance on his or her own judgment, as opposed to the stakeholder judgments. Erkanli, Stangl, and Muller (1993) were the first to apply BDE to estimation using ordinal inputs. However, they first convert the rank orderings to cardinal values (a process that may introduce additional information and biases) and then treat those cardinal values as if they were independent (even though they could not be, since the underlying ordinal rankings clearly could not be independent). Wang and Bier (2013) avoid these pitfalls by treating the entire set of rank orderings from a given stakeholder as a single observation (thus inherently accounting for the lack of independence between rank orderings given by the same stakeholder) and using the rank orderings directly (rather than converting them into cardinal values first). In the next section, we briefly discuss the mathematical models of conjoint analysis, PI, and BDE, when applied to generate estimates for attribute weights from preference rankings provided by stakeholders.

ESTIMATING ATTRIBUTE WEIGHTS FROM ORDINAL PREFERENCE RANKINGS

Estimating attribute weights from preferences over alternatives requires that the stakeholder(s) be presented with a set of N alternatives to compare and give the top R rankings (when $R=N$, the stakeholders give a complete ranking of the alternatives). The process also requires a pre-defined set of attributes and the single-attribute utilities of each alternative on each attribute (i.e., the values of v_{nm} for $n=1, \ldots, N$, and $m=1, 2$). For convenience, we denote the rank ordering of stakeholder k (for $k=1, \ldots, K$) by an ordered set of N distinct target indices $\mathrm{RO}^{(k)} = \left\{ n_1^{(k)}, \ldots, n_N^{(k)} \right\}$, where $n_r^{(k)}$ is the index of the rth most preferred alternative for $r=1, \ldots, R$; for $r=R+1, \ldots, N, n_r^{(k)}$ is the index of one of the $N-R$ unranked alternatives. The task of conjoint analysis, PI, or BDE is then to generate cardinal estimates (point estimates or probability distributions) for the various attribute weights w_m that best fit the stakeholders' rank orderings $RO^{(k)}$.

In particular, we focus on the LINMAP method of conjoint analysis, which generates only point estimates for the known attribute weights (w_1, w_2) using the standard additive multiattribute utility function without unobserved attributes. By contrast, PI and BDE allow for the possibility of an unobserved attribute and can generate probability distributions for the known attribute weights plus the weight of the unobserved attribute (w_1, w_2, w_3).

Conjoint Analysis: LINMAP

The LINMAP method of conjoint analysis is designed for eliciting the preferences of a single stakeholder. Therefore, for simplicity, in this section, we suppress the superscripts of $RO^{(k)}$ and $n_r^{(k)}$. For a given rank ordering RO, we can specify $(R(R-1)/2)+R(N-R)$ pairwise inequalities involving the utilities of the N alternatives; that is, $U_{n_i} > U_{n_j}$ for $1 \le i \le R$ and $i < j \le N$. According to Equation 5.1, given a set of estimated attribute weights $\hat{w} = (\hat{w}_1, \hat{w}_2)$ (without considering the unobserved attribute), the overall utility of alternative n can be estimated as

$$U_n(\hat{w}) = \hat{w}_1 v_{n1} + \hat{w}_2 v_{n2}.$$

For a given pair of alternatives i and j such that i is preferred to j according to the stated rank ordering RO (i.e., $n_i < n_j$), we then define the pairwise "violation" of i and j as $z_{ij}(\hat{w}) = \max\{0, U_j(\hat{w}) - U_i(\hat{w})\}$. LINMAP aims to find a set of weight estimates $\hat{w}$ that yields the smallest sum of pairwise violations from the rank ordering RO. In particular, the optimization problem is given by

$$\min_{\hat{w},z} \sum_{1 \le i \le R} \sum_{i < j \le N} z_{n_i n_j} \tag{5.3}$$

$$U_{n_i}(\hat{w}) - U_{n_j}(\hat{w}) + z_{n_i n_j} \ge 0 \text{ for } 1 \le i \le R; i < j \le N$$

$$z_{n_i n_j} \ge 0 \text{ for } 1 \le i, j \le N$$

$$\hat{w}_1 + \hat{w}_2 = 1; \text{ and } \hat{w}_m \ge 0 \text{ for } m = 1, 2.$$

This linear programming problem may generate nonunique optimal attribute weights, especially when there exist weight estimates $(\hat{w}_1, \hat{w}_2)$ that can perfectly reproduce RO. For purposes of this chapter, however, we simply pick an arbitrary set of optimal weights if there are multiple solutions.

The literature recommends applying LINMAP to each individual stakeholder separately, but is not clear on how to aggregate the resulting weight estimates across multiple stakeholders. Therefore, in this chapter, we use a simple ad hoc aggregation rule (Shocker and Srinivasan, 1979)—namely, averaging the point estimates of the attribute weights obtained from the various stakeholders.

Probabilistic Inversion

The task of PI as applied in Wang and Bier (2013) is to find a joint probability distribution for the uncertain attribute weights (including the weight for the unobserved attribute) $w = (w_1, w_2, w_3)$ and the unobserved attribute utilities $y = (y_1, \ldots, y_N)$ as given in 5.2, such that the distribution over (w, y) can reproduce the empirical distribution of the stakeholders' rank orderings. In particular, the empirical distribution of the rankings given by K stakeholders is specified by a matrix P, where element P_{rn} represents the probability that alternative n is ranked at the rth place by a randomly chosen one of the K stakeholders (for $r = 1, \ldots, R$ and $n = 1, \ldots, N$). For example, if $P_{12} = 0.5$, it means that 50% of the stakeholders rank alternative 2 at the first place.

In fact, multiple distributions for (w, y) can generate rankings whose marginal distribution matches the empirical distribution P. However, PI aims to select the matching distribution that is "closest" (in the sense of Kullback–Leibler divergence) to a predefined starting distribution Q_0. In this chapter, we assume that Q_0 is noninformative—in particular, we choose a "flat" Q_0 by assigning a Dirichlet distribution over (w_1, w_2, w_3) with all parameters equal to one and independent uniform distributions on [0, 1] for each of the y_n ($n = 1, \ldots, N$). If desired, however, we could also consider other types of noninformative starting distributions (e.g., U-shaped instead of uniform) or an informative starting distribution.

A Monte Carlo simulation-based implementation of PI then proceeds as follows. First simulate S independent samples $(w^{(s)}, y^{(s)})$ (for $s = 1, \ldots, S$) from the starting distribution Q_0. Then find the optimal discrete distribution $q = (q_1, \ldots, q_S)$ over the S samples by solving the following convex optimization problem:

$$\min_{q \in \Delta_S(1)} \sum_{s=1}^{S} q_s \ln\left(Sq_s\right) \tag{5.4}$$

$$\sum_{s=1}^{S} q_s J_{rn}\left(w^{(s)}, y^{(s)}\right) = P_{rn} \text{ for } r = 1, \ldots, R; n = 1, \ldots, N,$$

where $\Delta_S(1)$ is the simplex defined by $\left\{ q \in R_+^S \middle| \sum_{s=1}^{S} q_s = 1 \right\}$, and $J(w^{(s)}, y^{(s)})$ is an R-by-N indicator matrix. In particular, for a given set of $w^{(s)}$ and $y^{(s)}$, the function $J_{rn}(w^{(s)}, y^{(s)})$ equals 1 if alternative n is ranked in the rth place and 0 otherwise. Note that the inclusion of the unobserved attribute ensures the existence of feasible solutions in PI; see more details in Wang and Bier (2013).

Bayesian Density Estimation

The idea of BDE as applied in Wang and Bier (2013) is to update the decision-maker's prior distribution over (w, y) to yield a posterior distribution, by treating each stakeholder's rank ordering of the alternatives as an independent observation. In particular, we call the set of all possible values of (w, y) that are consistent with stakeholder k's rank ordering $RO^{(k)}$ the "active region" $AR^{(k)}$ for stakeholder k (for $k = 1, \ldots, K$); that is,

$$AR^{(k)} = (w, y) \in \Delta_3(1) \times [0,1]^N \text{ such that}$$
$$U_{n_i^{(k)}}(w, y) \geq U_{n_j^{(k)}}(w, y) \text{ for } 1 \leq i \leq R \text{ and } i < j \leq N, \tag{5.5}$$

where $U_n(w, y)$ is the multiattribute utility function as given in Equation 5.2.

If we assume that the decision-maker's prior distribution over the uncertain parameters (w, y) is randomly chosen according to a Dirichlet process (Ferguson, 1973) concentrated at Q_0 with self-trust degree $\alpha > 0$, then the posterior distribution for (w, y) is also random and follows a Dirichlet process. In particular, the expected posterior distribution for (w, y) after observing K stakeholders' rank orderings is given by

$$\frac{\alpha}{\alpha + K} Q_0 + \frac{1}{\alpha + K} \sum_{k=1}^{K} Q_0^{(k)}, \tag{5.6}$$

where $Q_0^{(k)}$ is a truncated version of the starting distribution Q_0 that puts nonzero mass on only points inside AR$^{(k)}$; that is, $Q_0^{(k)} \sim Q_0 \cdot 1_{\{(w,y)\in AR^{(k)}\}}$ with $1_{\{\cdot\}}$ being the indicator function. Note that Equation 5.6 suggests a linear pooling of the decision-maker's prior knowledge with the stake-holders' judgments, where the value of α (relative to the number of stakeholders K) reflects the level of reliance on the decision-maker's prior belief.

Statistical estimates for the moments of the distribution 5.6 can be obtained using Monte Carlo simulation (e.g., Gibbs sampling). In particular, to draw S random samples from 5.6, we need to randomly generate $S\alpha/(\alpha+K)$ points from the entire sample space of (w, y) and $S/(\alpha+K)$ points from each of the active regions AR$^{(k)}$ for the K stake-holders, both according to the predefined starting distribution Q_0. (Note that the value of S must be chosen so that the sample sizes $S/(\alpha+K)$ and $S\alpha/(\alpha+K)$ are integers.) Again, more details of this methodology are available in Wang and Bier (2013).

Relationship between LINMAP, PI, and BDE

In general, LINMAP is designed for elicitation of the preferences of a single stakeholder, while PI and BDE can be used for multiple stake-holders. LINMAP also generates only point estimates for the attribute weights, while PI and BDE produce probability distributions. However, there are some other interesting relationships between LINMAP, PI, and BDE.

First, consider the case when there is only one stakeholder (the default setting for the application of LINMAP) or when all stakeholders give the same rank ordering. Wang and Bier (2013) show that if zero weight is placed on the decision-maker's judgment (i.e., the self-trust degree $\alpha \to 0$), then PI and BDE are equivalent in this case.

When there is only one stakeholder and his/her rank ordering of alternatives can be perfectly explained by the available set of attrib-utes, then LINMAP achieves zero pairwise violations (possibly with multiple optimal solutions). BDE (or PI) is also feasible in this case even if we exclude the unobserved attribute. In fact, each point in the active region of BDE corresponding to the stakeholder's stated rank ordering (without considering the unobserved attribute) is one pos-sible set of optimal attribute weights that could be produced by LINMAP.

If the stakeholder's given rank ordering of alternatives does not deviate too much from the available set of attributes, then LINMAP can still obtain a set of estimated attribute weights that fits the stakeholder's judgments reasonably well. Both PI and BDE would need to include the unobserved attribute to ensure feasibility in this case but would presumably place small weight on it. Moreover, we would anticipate the LINMAP estimates to be quite close to the means of the probabilistic weights generated by PI or BDE.

However, if the stakeholder's rank ordering of alternatives is based heavily on factors other than the available set of attributes, then any possible weights that do not consider the unobserved attribute might lead to substantial deviations from the stated ordinal ranking. Therefore, the LINMAP estimates are expected to perform poorly in reproducing the stated rank ordering in this case, while PI (or BDE) would perform better but would assign high weight to the unobserved attribute.

We now consider the case when there are multiple stakeholders who provide different ordinal judgments. PI and BDE will in general give different results in this case. Consider multiple groups of stakeholders with different sets of ordinal rankings, where each group nonetheless yields the same empirical marginal ranking distribution P. PI would give the same result for all of these groups of stakeholders, while BDE would in general give different results for groups with different rank orderings. However, Wang and Bier (2013) show that if we choose a flat starting distribution Q_0, then the PI result will coincide with the BDE result with the highest entropy.

The fact that PI exploits only marginal stakeholder rankings may make it suitable to use when the available judgments are not highly reliable (e.g., when using only a small number of stakeholders to represent the opinions of a much larger population), in which case we may not want to put too much emphasis on the empirical correlations between stakeholders. We could also use BDE in that case with a large self-trust degree assigned to the decision-maker's prior knowledge.

However, when rankings from large numbers of stakeholders are available, we may wish to explicitly account for correlated rank orderings (e.g., if stakeholders who rank alternative 1 higher than 2 also rank alternative 3 higher than 4 and vice versa). PI is not able to capture the effects of such correlations, since it uses only the empirical marginal ranking distribution P. By contrast, BDE is sensitive to correlations among the rankings and thus will be more appropriate if we want to account for the existence of different "schools of thought" (i.e., sub-groups of stakeholders who hold similar opinions). The results in Wang and Bier (2013) also suggest that BDE tends to give multimodal distributions more often than PI in the face of multiple schools of thought. Therefore, when sufficient numbers of stakeholder judgments have been obtained to provide reasonable assurance that any apparent schools of thought are meaningful, BDE might be more appropriate.

ILLUSTRATIVE CASE STUDY

In this section, we present a case study on elicitation of adversary preferences about attacks against major US urban areas, to illustrate the applicability of LINMAP, PI, and BDE. In particular, we consider the 10 urban areas with the highest expected damage from terrorism (according to Willis et al., 2005), including New York City (NYC); Chicago; San Francisco; Washington, DC; Los Angeles (LA); Philadelphia; Boston; Houston; Newark; and Seattle. We consider two known attacker attributes, expected yearly property loss from terrorism and population density (Willis et al., 2005), plus an unobserved attribute. The original values of each known attacker attribute m ($m = 1, 2$) for each city n ($n = 1, \ldots, 10$) are presented in Table 5.1. In order to apply LINMAP, PI, and BDE, we normalize these values to range between 0 and 1 (as shown by v_1 and v_2 in Table 5.1). We assume that two hypothetical intelligence experts have given the rankings of the top five most attractive cities in Table 5.2.

We first apply LINMAP to get point estimates for the two known attribute weights from the rank orderings of the two experts separately. In the two-expert case, we average the LINMAP weights to get aggregate estimates. We then use PI and BDE to get probability distributions (and expected values) for the two known attribute weights plus the weight of the unobserved attribute, for the two experts separately and for both experts together. We choose a flat starting distribution for both PI and BDE. For comparison purposes, we set the self-trust degree $\alpha \to 0$ in BDE since PI does not have such a parameter. (In that case, PI and BDE give the same results when there is only one expert.) See results in Figure 5.1.

Note that the hypothetical judgments of expert 1 are well represented by the two known attributes and are more closely related to

Table 5.1   Attribute values for the 10 US cities with the highest expected terrorism losses

	Property loss		Population density	
	$ million	v_1	per sq mile	v_2
New York City	413	1.000	8159	1.000
Chicago	115	0.278	1634	0.200
San Francisco	57	0.138	1705	0.209
Washington DC	36	0.087	756	0.093
Los Angeles	34	0.082	2344	0.287
Philadelphia	21	0.051	1323	0.162
Boston	18	0.044	1685	0.207
Houston	11	0.027	706	0.087
Newark	7.3	0.018	1289	0.158
Seattle	6.7	0.016	546	0.067

Table 5.2   Hypothetical expert rank orderings for the illustrative case study

Rank	Expert 1	Expert 2
1	New York City	District of Columbia
2	Chicago	Chicago
3	Los Angeles	New York City
4	San Francisco	San Francisco
5	Philadelphia	Houston

property loss than to population density. The mean elicited weights for both PI and BDE reflect this by generating mean weights of $E[w_1] = 0.44$ for property loss and $E[w_2] = 0.34$ for population density and a small weight of $E[w_3] = 0.22$ for the unobserved attribute (as shown in the first row of Fig. 5.1). LINMAP omits the unobserved attribute but obtains qualitatively similar results, placing higher weight on property loss ($\hat{w}_1 = 0.59$) than on population density ($\hat{w}_2 = 0.41$).

However, LINMAP behaves quite differently from either PI or BDE when using the hypothetical judgments of expert 2 (see the second row of Fig. 5.1). This is because the hypothetical rank orderings of expert 2 were designed to reflect a preference for high property loss but

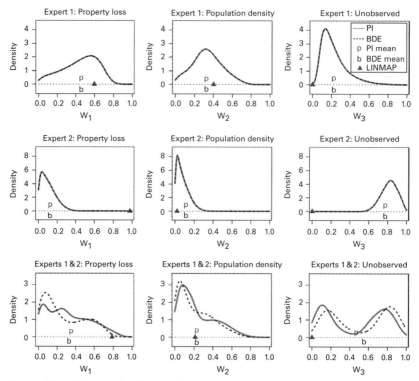

Figure 5.1   Elicited probability densities and point estimates for the attribute weights.

some aversion to attacks on densely populated cities (such as NYC). For example, these rankings might represent the target preferences of an attacker who wishes to cause significant economic damage and make a big splash but without large numbers of fatalities. Therefore, if the weight on population density is constrained to be positive, the two known attributes are inadequate to fully capture the judgments of expert 2. The best fit obtained by LINMAP assigns an extremely high weight to property loss ($\hat{w}_1 = 0.98$), but this assignment cannot explain, for example, why expert 2 ranks DC as the most attractive city and Houston among the top five, since both cities have relatively low property losses. By contrast, PI and BDE identify that there may be other factors driving the judgments of expert 2, by placing a high expected weight $E[w_3] = 0.82$ on the unobserved attribute.

When considering the judgments of both experts 1 and 2 combined, PI and BDE also seem to give more sensible results than LINMAP (see the third row of Fig. 5.1). In particular, both PI and BDE give multimodal distributions with high variance for the attribute weights, to explicitly account for the disagreement between the two experts. (In fact, BDE gives more noticeably multimodal results than PI for the weight assigned to property loss.) However, the average of the weights given by LINMAP camouflages such disagreement.

Thus, when an expert's judgments are well represented by the known attributes, LINMAP is able to find a set of estimated attribute weights that give a good fit to the expert's rank ordering of alternatives. Moreover, limited evidence in this chapter also suggests that in this case the LINMAP estimates will be reasonably close to the mean weights obtained using PI or BDE. However, when the known attributes do not adequately explain the expert judgments, LINMAP might generate nonsensical results, while PI and BDE can increase the weight on the unobserved attribute to reflect the fact that the available set of attributes is incapable of fully expressing the expert preferences.

ALLOWING FOR NEGATIVE WEIGHTS

Of course, stakeholders may disagree about whether a larger value of a particular attribute yields higher or lower utility. For example, in the context of adversarial preference modeling, some intelligence experts may believe that a particular adversary group favors smaller rather than larger numbers of fatalities, since attacks with a large number of fatalities could lead to reduced support for the terrorist's cause and/or massive US retaliation. However, other experts may believe that the adversary would prefer attacks that cause more fatalities, in order to evoke more fear.

To model this, CREATE (2011) proposes extending the standard additive multiattribute utility function to allow for negative weights while restricting the sum of the absolute values of the various attribute weights to equal 1 (i.e., $|\hat{w}_1| + |\hat{w}_2| = 1$ for LINMAP and $|\hat{w}_1| + |\hat{w}_2| + |\hat{w}_3| = 1$ for PI or BDE). This choice of attribute weights allows enough flexibility to deal with stakeholder disagreement on the directions of the attribute weights but still constrains their magnitudes.

Note that for PI and BDE, we still require that the weight for the unobserved attribute be nonnegative (i.e., $w_3 \geq 0$). There are two reasons for this choice. First, the fact that the method (PI or BDE) fits not only

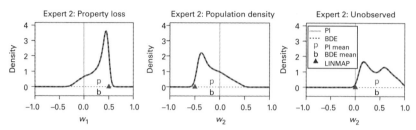

Figure 5.2 Elicited probability densities and point estimates for attribute weights when allowing negative weights.

the weight w_3 but also the values of the unobserved attribute y_n for the various alternatives implies that no degrees of freedom are lost by restricting w_3 to be positive. This is because the attribute values can still be chosen appropriately to reflect stakeholders' disagreement on the directions of the attribute scales. Second, if w_3 is allowed to be negative, then the expected value $E[w_3]$ will no longer capture the full effect of the unobserved attribute, complicating the interpretation of the results.

In the illustrative case study of the previous section, when applying LINMAP, PI, and BDE with nonnegative weights to the rank ordering given by expert 2, we find that population density seems to be almost totally unimportant to the attacker ($\hat{w}_2 = 0.02$ for LINMAP and $E[w_2]=0.06$ for PI or BDE, as shown in the second row of Fig. 5.1). However, by allowing for negative attribute weights, the LINMAP weight for population density becomes significantly negative ($\hat{w}_2 = -0.5$). The PI and BDE results also give a probability of 0.73 of a negative weight on population density (as shown in Fig. 5.2), with mean $E[w_2]=-0.18$. This reflects the observation that the judgments of expert 2 would be consistent with an adversary that is averse to attacks on cities with high population densities. Interestingly, although the LINMAP weights ($\hat{w}_1 = 0.5$; $\hat{w}_2 =-0.5$) are not close to the mean weights elicited by PI or BDE ($E[w_1]=0.26$; $E[w_2]=-0.18$), they are quite close to the modes of the distributions for w_1 and w_2 given by PI and BDE.

Thus, elicitation methods with only nonnegative weights may incorrectly suggest that a given attribute is irrelevant or unimportant. Allowing for negative weights avoids this pitfall and also automatically accounts for the possibility of conflicting views among the stakeholders on whether a larger value of an attribute corresponds to higher or lower utility. This automated assessment is especially important for elicitation tasks with large numbers of stakeholders (e.g., using online surveys), because it avoids the need to interview individual stakeholders about their (possibly differing) perceptions of the various attribute scales.

RELIABILITY OF PARTIAL RANK ORDERINGS

The elicitation methods discussed here are practical only if stakeholders can provide partial rank orderings of the available alternatives, rather than a full ranking of all alternatives, which may become cumbersome. In fact, there would be little reason to even obtain estimates of attribute weights, if stakeholders had to provide direct rankings of all alternatives in order to generate those estimates. Therefore, in this section, we conduct simulation-based sensitivity analysis to investigate whether elicitation results from partial rank orderings of alternatives are sufficiently reliable for use in practice.

For simplicity, we apply LINMAP and BDE to the case of a single stakeholder, without allowing for negative attribute weights. We also choose a flat starting distribution for BDE and set the self-trust degree $\alpha \to 0$. (Note again that when there is only one expert, PI and BDE give the same results.)

We consider three main factors that might affect the reliability of partial rank orderings. First, for a fixed number of alternatives, we would expect that ranking more of the alternatives (i.e., larger R) would generate more reliable results than ranking fewer alternatives. However, it is important to know how quickly results become reliable as the number of ranked alternatives increases.

Second, the total number of alternatives (N) also seems to be an important factor. However, we have no prior hypothesis on whether results would be more reliable for a larger or smaller total number of alternatives.

Finally, we are interested in how the number of known attributes (M) affects the reliability of partial rank orderings. Partial rankings should hypothetically be less reliable for estimating large numbers of attribute weights, since in that case there are simply more parameters to estimate. Moreover, as the number of attributes grows, the various attributes would be more likely to be correlated, possibly leading to less stable elicitation results. Therefore, when the total number of attributes is greater, we expect that results derived from partial rank orderings would deviate more from those based on full rank orderings.

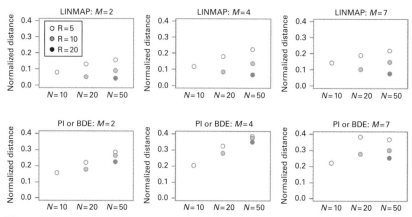

Figure 5.3   Normalized distance between estimated weights using full rankings versus partial rankings.

In particular, we consider three levels for the total number of alternatives ($N = 10$, 20, or 50) and three levels for the number of known attributes ($M = 2$, 4, or 7). For each number of alternatives N, we compare different levels of partial ranking R against full ranking of all N alternatives. When $N = 10$, we consider one partial rank ordering identifying the top $R = 5$ alternatives; when $N = 20$, we consider two partial rank orderings, the top $R = 5$ or 10 alternatives; when $N = 50$, we consider three cases, the top $R = 5$, 10, or 20 alternatives.

Here, reliability is measured by the Euclidean distance between the (mean) attribute weights elicited using full rank orderings and those obtained from partial rank orderings. For ease of interpretation, we normalize this distance by dividing it by the square root of 2, which is the largest possible Euclidean distance between two nonnegative weights that sum up to unity, to yield a maximum normalized distance of 1. (Of course, the ideal distance is 0.)

For each of the aforementioned cases, we randomly generate 400 sets of attribute values v_{nm} and rank orderings of the attractiveness of alternatives U_n. Four hundred runs are sufficient to ensure that the simulation error is within $\pm 10\%$ of the quantity of interest. Figure 5.3 presents the estimated normalized distances between the attribute weights elicited using partial versus full rank orderings in our simulations.

In general, larger subsets of partial rankings (i.e., larger R) lead to more reliable estimates for the attribute weights (smaller normalized distances). This is true for all three methods (LINMAP, PI, and BDE), at least for the cases we considered.

Moreover, ranking roughly 20% of the alternatives (e.g., the top 5 of 20 or the top 10 of 50) seems to provide sufficiently good weight estimates for the use of LINMAP, with a normalized distance of less than 0.2 from the weight estimates generated by full rankings. However, more alternatives need to be ranked for PI or BDE to achieve an equivalent level of reliability. This seems reasonable, because in order to generate probability distributions for the weights rather than only point estimates, PI and BDE attempt to get more information out of the data than LINMAP and thus may need more data in order to be reliable.

In spite of the need for larger numbers of partial rankings, PI and BDE are still not impractical. In fact, ranking roughly 50% of the alternatives (e.g., the top 5 of 10, the top 10 of 20, or the top 20 of 50) yields moderately reliable results (with a normalized distance of <0.3) in most cases.

We now compare the results given by ranking the top 50% of all alternatives for different numbers of alternatives (N), again from Figure 5.3. Larger numbers of alternatives (i.e., larger N) seem to give better results for LINMAP, but not for PI or BDE. This discrepancy makes the use of these methods suitable in different circumstances. In particular, LINMAP may perform better than PI and BDE for problems with large numbers of alternatives.

Finally, we investigate the effect of varying the number of known attributes (M). All three methods seem to work better for smaller numbers of attributes (i.e., smaller M). For example, when there are only two known attributes ($M = 2$), asking the stakeholders to provide only the top 5 out of 50 alternatives (i.e., the top 10%) seems to produce quite reliable estimates for the attribute weights using LINMAP, PI, or BDE.

CONCLUSIONS AND DIRECTIONS FOR FUTURE RESEARCH

This chapter focuses on an elicitation process that uses (partial) rank orderings of a given set of alternatives and mathematically derives estimates for the various attribute weights in an additive multiattribute utility function that represents the stakeholders' preferences.

The LINMAP method in conjoint analysis has been popular since its emergence in the 1970s. It is simple and fast and can yield good point estimates for the attribute weights when a stakeholder's rank orderings of alternatives do not deviate too much from the available set of known attributes. On the other hand, PI and BDE have only recently been applied to elicitation of ordinal judgments. Both of these newer methods can provide probability distributions over the attribute weights, thus explicitly accounting for the effects of uncertainty and disagreement in group decision-making. PI and BDE can also generate weights for unobserved attributes, which is useful in quantifying whether the available set of attributes are adequate to capture the stakeholders' rank orderings.

This chapter discusses extensions of these methods to allow for negative attribute weights (CREATE, 2011). Some attributes that seem totally unimportant when using only nonnegative weights may turn out to be surprisingly important after we drop the nonnegativity restriction. This is significant, since some stakeholders may disagree with the pre-defined scales of the attributes—for example, thinking that a larger value of a given attribute corresponds to lower utility. Allowing for negative weights handles that possibility in an automatic and realistic way. This feature makes these methods especially useful for large-scale elicitations (e.g., through online surveys).

This chapter also examines the reliability of using partial rank orderings in eliciting attribute weights, compared to full rank orderings. The reliability of the methods appears to depend on the total number of alternatives, the number of ranked alternatives, and the number of known attributes. In general, ranking only 20% of the alternatives seems to be sufficient for the use of LINMAP, but more rankings (perhaps 50%) may be needed for the use of PI or BDE.

The elicitation methods presented in this chapter have all been developed based on the assumption that stakeholders give rank orderings without ties. Shocker and Srinivasan (1974) have extended LINMAP to allow for tied rankings, by treating tied alternatives as having utilities that are "close" to each other. We could also extend PI and BDE to accommodate tied rankings in a similar way. However, further research is needed to investigate the performance of LINMAP, PI, and BDE in the face of tied rank orderings.

In other work, we have found that when the values of some attributes are highly correlated with each other (which will tend to happen quite often when the number of attributes is large), the elicitation results given by both PI and BDE can be unstable in the face of small changes in the attribute values. This effect is analogous to the problem of collinearity in multiple regression. Therefore, an interesting topic for future research would be to develop a "significance test" to assess whether removing an attribute would significantly change the relative weights of the remaining attributes as well as the importance of the unobserved attribute.

Still, the basic approach of ordinal analysis discussed in this chapter is a rigorous and practical method and can readily be applied—for example, as a source of input for convergence validation.

ACKNOWLEDGMENTS

This research was supported by the US Department of Homeland Security through the National Center for Risk and Economic Analysis of Terrorism Events under Grant 2010-ST-061-RE0001. However, any opinions, findings, and conclusions or recommendations in this document are those of the authors and do not necessarily reflect views of the US Department of Homeland Security.

REFERENCES

Abbas, A. E. 2006. Maximum entropy utility. Operations Research **52**(2) 277–290.

Ali, S., S. Ronaldson. 2012. Ordinal preference elicitation methods in health economics and health services research: Using discrete choice experiments and ranking methods. British Medical Bulletin **103** 21–44.

Barron, F. H., B. E. Barrett. 1996. The efficacy of SMARTER—Simple multi-attribute rating technique extended to ranking. Acta Psychologica **93** 23–36.

Bedford, T. J., R. M. Cooke. 2001. Probabilistic Risk Analysis: Foundations and Methods. Cambridge University Press, New York/Cambridge.

Borcherding, K., T. Eppel, D. von Winterfeldt. 1991. Comparison of weighting judgements in multiattribute utility measurement. Management Science **37**(12) 1603–1619.

Center for Risk and Economic Analysis of Terrorism Events. 2011. Adaptive Adversary Modeling for Terrorism Risk Analysis: Final Report, University of Southern California, Los Angeles, CA.

Clemen, R. T., T. Reilly. 2001. Making Hard Decisions with Decision Tools. Duxbury, Belmont, CA.

Cooke, R. M. 1994. Parameter fitting for uncertain models: Modeling uncertainty in small models. Reliability and Engineering System Safety **44** 89–102.

Dawes, R. M., B. Corrigan. 1974. Linear models in decision making. Psychological Bulletin **81** 95–106.

Edwards, W. 1977. How to use multiattribute utility measurement for social decision making. IEEE Transactions on Man, Systems, and Cybernetics **7** 326–340.

Edwards, W., F. H. Barron. 1994. SMARTS and SMARTER: Improved simple methods for multiattribute utility measurement. Organizational Behavior and Human Decision Processes **60** 306–325.

Erkanli, A., D. K. Stangl, P. Muller. 1993. A Bayesian analysis of ordinal data. ISDS Discussion, Paper 93-A01, Duke University, Durham, NC.

Fechner, G. T. 1860. Elemente der Psychophysik (Elements of Psychophysics). Breitkopf und Härte, Leipzig.

Ferguson, T. S. 1973. A Bayesian analysis of some nonparametric problems. The Annals of Statistics **1** 209–230.

Green, P. E., V. Srinivasan. 1978. Conjoint analysis in consumer research: Issues and outlook. Journal of Consumer Research **5**(2) 103–123.

Green, P. E., A. M. Krieger, Y. Wind. 2001. Thirty years of conjoint analysis: Reflections and prospects. Interfaces **31**(3) S56–S73.

Horsky, D., M. R. Rao. 1984. Estimation of attribute weights from preference comparisons. Management Science **30**(7) 801–822.

Jenelius, E., J. Westin, Å. J. Holmgren. 2010. Critical infrastructure protection under imperfect attacker perception. International Journal of Critical Infrastructure Protection **3**(1) 16–26.

Keeney, R. L. 2007. Developing objectives and attributes. W. Edwards, R. F. Miles, D. von Winterfeldt, eds. Advances in Decision Analysis: From Foundations to Applications. Cambridge University Press, New York, 104–128.

Keeney, R. L., H. Raiffa. 1976. Decisions with Multiple Objectives: Preferences and Value Trade-Offs. John Wiley & Sons, Inc., New York.

Keeney, R. L., D. von Winterfeldt. 1994. Managing nuclear waste from power plants. Risk Analysis **14** 107–130.

Kirkwood, C. W., R. K. Sarin. 1985. Ranking with partial information: A method and an application. Operations Research **33**(1) 38–48.

Kraan, B., T. Bedford. 2005. Probabilistic inversion of expert judgments in the quantification of model uncertainty. Management Science **51**(6) 995–1006.

McCabe, C., J. Brazier, P. Gilks, A. Tsuchiya, J. Roberts, A. O'Hagan, K. Stevens. 2006. Using rank data to estimate health state utility. Journal of Health Economics **25** 418–431.

McFadden, D. 1977. Quantal choice analysis: A survey. Annals of Economic and Social Measurement **5** 363–390.

McFadden, D., K. Train. 2000. Mixed MNL models for discrete response. Applied Econometrics **15** 447–470.

Neslo, R., F. Micheli, C. V. Kappel, K. A. Selkoe, B. S. Halpern, R. M. Cooke. 2011. Modeling stakeholder preferences with probabilistic inversion: Application to prioritizing marine ecosystem vulnerabilities. I. Linkov, E. Ferguson, V. Magar, eds. Real Time and Deliberative Decision Making: Application to Risk Assessment for Nonchemical Stressors. Springer, Amsterdam, NL, 265–284.

Rao, J. S., M. Sobel. 1980. Incomplete Dirichlet integrals with applications to ordered uniform spacings. Journal of Multivariate Analysis **10** 603–610.

Revelt, D., K. Train. 1998. Mixed logit with repeated choices: Households' choice of appliance efficiency level. Review of Economics and Statistics **LXXX**(4) 647–657.

Sarabando, P., L. C. Dias. 2009. Multiattribute choice with ordinal information: A comparison of different decision rules. IEEE Transactions on Systems, Man and Cybernetics, Part A: Systems and Humans **39**(3) 545–554.

Schoemaker, P. J. H., C. D. Waid. 1982. An experimental comparison of different approaches to determining weights in additive utility models. Management Science **28** 182–196.

Shocker, A. D., V. Srinivasan. 1973. Linear programming techniques for multi-dimensional analysis of preferences. Psychometrika **38**(6) 337–369.

Shocker, A. D., V. Srinivasan. 1974. A consumer-based methodology for the identification of new product ideas. Management Science **20**(6) 921–937.

Shocker, A. D., V. Srinivasan. 1979. Multiattribute approach for product concept evaluation and generation: A critical review. Journal of Marketing Research **16**(2) 159–180.

Stewart, T. J. 1996. Robustness of additive value function methods in MCDM. Journal of Multi-Criteria Decision Analysis **5** 301–309.

Stillwell, W. G., D. A. Seaver, W. Edwards. 1981. A comparison of weight approximation techniques in multiattribute utility decision making. Organizational Behavior and Human Performance **28** 62–77.

von Winterfeldt, D., W. Edwards. 1986. Decision Analysis and Behavioral Research. Cambridge University Press, New York.

Wang, C., V. M. Bier. 2013. Expert elicitation of adversary preferences using ordinal judgments. Operations Research **61**(2) 372–385.

Watson, S. R., D. M. Buede. 1987. Decision Synthesis. Cambridge University Press, Cambridge.

Willis, H. H., A. R. Morral, T. K. Kelly, J. J. Medby. 2005. Estimating Terrorism Risk. RAND Corporation, Santa Monica, CA.

CHAPTER 6

Causal Analysis and Modeling for Decision and Risk Analysis

Louis Anthony (Tony) Cox, Jr.
Cox Associates, NextHealth Technologies,
University of Colorado-Denver, Denver, CO, USA

INTRODUCTION: THE CHALLENGE OF CAUSAL INFERENCE IN RISK ANALYSIS

In decision and risk analysis, the goal of making decisions is to change the probabilities of outcomes to make preferred outcomes more likely. The decision-maker's choices act upon the world, causing changes in outcome probabilities. Yet, this crucial concept of causal efficacy is seldom developed in detail in decision analysis, and the fact that formal probability theory applies only to events (subsets of a sample space) rather than to actions and their consequences is seldom emphasized (Pearl, 2010). Yet, if one is careless about causation—for example, by routinely interpreting statistical associations or regression coefficients as if they were known to be causal, as is unfortunately common practice in modern epidemiology and public health applications—then the decisions that one takes based on assumptions about causality may turn out to cause quite different shifts in outcome probabilities than those

Breakthroughs in Decision Science and Risk Analysis, First Edition.
Edited by Louis Anthony Cox, Jr.

that were expected and intended. Fortunately, modern methods of causal analysis can do much to prevent such unpleasant surprises. This chapter surveys some of the most useful methods for causal analysis, contrasts them with widely used but untrustworthy methods based on judgments about statistical associations, and develops their relations to Bayesian networks (BNs) and influence diagrams (IDs).

Our starting point is recognition that public health risk managers, regulators, policy makers, and others in positions of political influence or power are frequently presented with conflicting accounts of how the world works. They are urged by various interest groups—often passionately—to take different prompt, decisive actions based on these rival causal theories. Members of the Congress are implored by many climate scientists to do more to curb climate change before it is too late. Simultaneously, other groups beseech them not to spend resources on expensive actions that might create no, or little, or uncertain, benefits. While many financial economists and risk analysts call for tighter regulation of complex financial instruments, or better-funded public safety nets for big banks, or quicker and larger stimulus expenditures, others warn that these efforts risk exacerbating the problems they are meant to solve. Experts in development economics are split between those who encourage increasing aid payments to poor countries to jump-start their economies and those who say that such transfers merely cement the wealth, power, and corruption of existing power elites, thus helping to keep their countries poor.

In these and countless other disagreements, both sides usually have more-or-less plausible stories about how their recommended actions will cause desirable consequences, but their stories do not agree. This puts risk managers, decision-makers, and policy makers in the uncomfortable position of having to assess the credibility of rival causal theories—a task for which compelling data, decisive expertise, and provably useful training are often in short supply. The task is complicated by the notorious difficulty of defining causality and by the fact that controlled experiments—or even careful replication of previous observations—are often impossible. Analysts and decision-makers must decide what to do based on observations of what has happened, without knowing what would have happened had different choices been made; history does not reveal its alternatives. Moreover, like risk itself, causality is often easier to model mathematically than to define well in words. Risk equations and causal models typically describe how changes in inputs to a system would change probabilities of outputs.

This provides the essential scientific information needed to inform risk management decisions. But such causal models and relations seldom lend themselves to clean verbal definitions of causality or of risk: they simply predict the changes in probable outcomes that would be caused by changes in inputs.

Example: Causality is Often Easier to Model Quantitatively than to Define Qualitatively

If water is left flowing into an initially empty bucket of volume V liters at a rate of r liters per minute for $t > V/r$ minutes, causing the bucket to overflow, then philosophers and legal scholars might ponder whether the overflow was caused by an inadequate design decision (V too small) or excessively stressful operating conditions (r too large) or careless risk management (t too large), but a modeler can simply predict that the bucket will overflow if and only if $t > V/r$ and leave to others the challenges of expressing this predictive conditional relation in satisfying causal terms. Even if the correct description of the problem parameters is uncertain, quantifying probabilities of an undesired outcome (overflow) from a joint probability distribution for (V, r, t), would not change or obscure the underlying causal relation. Neither would interconnecting multiple (possibly leaky) buckets or allowing flow rate r to be described by a stochastic process raise any conceptual difficulties for understanding the causal relations involved in calculating the probable time until the levels in one or more buckets first move outside a desired set of values. Many tasks in applied risk assessment involve such clear and easily modeled causal relations. Examples include predicting financial ruin probabilities (where money, rather than water, flows into and out of a business or an investment fund); developing biologically based cancer risk assessment models (where cell populations make random transitions among stages); protecting and managing ecosystems (where stocks of vulnerable species increase or decrease over time); operating complex engineering or industrial systems (where components may degrade and be inspected, maintained, replaced, or repaired over time); and performing microbial risk assessments (in which the burden of foodborne illnesses in a population changes as microbial safety practices change along the food supply chain). Risk models in these and other areas embed clear concepts of causation, often based on submodels or empirical estimates of how interventions change the rates of flows or transitions among adjacent compartments. Decisions about how to intervene to cause more desirable patterns of flows or transitions can then be based on the resulting risk models.

Two natural reactions to the challenge of judging among rival causal theories are to trust one's common sense and intuition, deferring to gut feel when cognition must admit defeat; and to rely on trusted scientific experts, who specialize in the relevant technical disciplines, for candid advice about the probable consequences caused by different choices. But modern scholarship has diminished the luster and apparent trustworthiness of both intuitive and expert judgments in matters of causation. As discussed further in Chapters 2 and 10, psychologists have shown convincingly that all of us, including experts in science and statistics, are prone to heuristics and biases that limit the trustworthiness of our judgments. These include overconfidence in our own judgments, misattribution of effects to causes, excessive inclination to blame people instead of situations, affect bias (in which emotional responses color our beliefs about facts, inclining us toward causal theories that agree with our intuitive perceptions of what is good or bad), motivated reasoning (which prompts us to believe whatever seems most profitable for us to believe), and confirmation bias (which leads us to see only what we expect and to seek and interpret information selectively to reinforce our beliefs rather than to learn from reality) (Fugelsang et al., 2004; Gardner, 2009; Sunstein, 2009).

For over a decade, the peer-reviewed scientific literature on risks and causes has been found to reflect these very human biases, with a large excess of false-positive errors in published results and in confident public assertions about health effects of various interventions (Ottenbacher, 1998; Imberger et al., 2011; Sarewitz, 2012). Attempts that fail to replicate published results may carry little professional or academic reward, undermining incentives to try to independently replicate key claims (Sarewitz, 2012; Yong, 2012). Scientists with deep subject matter expertise are not necessarily or usually also experts in causal analysis and valid causal interpretation of data, and their causal conclusions are often mistaken. This has led some commentators to worry that "science is failing us," due largely to widely publicized but false beliefs about causation (Lehrer, 2012), and that, in recent times, "Most published research findings are wrong" (Ioannidis, 2005), with the most sensational and publicized claims being most likely to be wrong. Yet, published research findings, and their supporting causal models, provide the essential bedrock of scientific and empirical support for decision processes based on analysis and deliberation. If this core scientific input is untrustworthy, then risk assessments and decisions based on it may also fail to perform as expected.

To feel the pull of rival causal theories, consider the contrasting accounts of public health effects caused by air pollution, shown in Table 6.1. On the left are quotes from studies usually interpreted as showing that exposure to air pollutants (mainly fine particulate matter (PM2.5)) causes increased risks of adverse health effects (e.g., Pope, 2010), along with some quantitative risk estimates for these effects. On the right are caveats and results of studies suggesting that these associations may not be causal after all. Both seem more or less plausible at first glance. Yet, important policy decisions, such as about whether the costs of reducing pollution levels further would be justified by resulting benefits from improved public health, depend crucially on which of these rival interpretations is correct.

If one's own judgment, scientific expert opinion, and the authority of peer-reviewed publications are all suspect as guides to the truth about such basic questions as whether air pollution caused adverse health effects in these studies, then how might one more objectively determine what causal conclusions are warranted by available facts and data, to serve as a basis for sounder risk-informed decision-making? A common approach in epidemiology is to use statistical tests to determine whether there is strong evidence for a nonrandom positive association between exposure and response and then to check whether, in the judgment of knowledgeable experts, the association can correctly be described by adjectives such as "strong," "consistent," "specific," "temporal," and "biologically plausible." The problem with this very popular approach is that all of these (and other) laudatory adjectives can apply perfectly well to associations even when there is no causation. Such associations can be created by strong confounders with time delays; or by data selection and model selection biases; or by unmodeled errors in exposure estimates; or by regression to the mean, or contemporaneous historical trends, or a host of other well-known threats to valid causal inference (Campbell and Stanley, 1966; Cox, 2007). Applying adjectives to associations, as proposed in the thoughtful and influential work of Sir Bradford Hill and as subsequently implemented in many weight-of-evidence schemes, does not overcome the basic limitation that an association is still only an association. Even the best qualified association

Table 6.1 Some examples of conflicting claims about health effects known to be caused by air pollution

Pro (causal interpretation or claim)	Con (counterinterpretation or claim)
"Epidemiological evidence is used to quantitatively relate PM2.5 exposure to risk of early death. We find that UK combustion emissions cause ~13,000 premature deaths in the UK per year, while an additional ~6000 deaths in the UK are caused by non-UK European Union (EU) combustion emissions" (Yim and Barrett, 2012)	"[A]lthough this sort of study can provide useful projections, its results are only estimates. In particular, although particulate matter has been associated with premature mortality in other studies, a definitive cause-and-effect link has not yet been demonstrated" (NHS, 2012)
"[A]bout 80,000 premature mortalities [per year] would be avoided by lowering PM2.5 levels to 5 $\mu g/m^3$ nationwide" in the US. 2005 levels of PM2.5 caused about 130,000 premature mortalities per year among people over age 29, with a simulation-based 95% confidence interval of 51,000–200,000 (Fann et al., 2012)	"Analysis assumes a causal relationship between PM exposure and premature mortality based on strong epidemiological evidence… However, epidemiological evidence alone cannot establish this causal link" (EPA, 2011, table 5.11)
	Significant negative associations have also been reported between exposures to some pollutants (e.g., NO2 (Kelly et al., 2011), PM2.5 (Krstić, 2011), and ozone (Powell, Lee, and Bowman, 2012)) and short-term mortality and morbidity rates.
"Some of the data on the impact of improved air quality on children's health are provided, including… the reduction in the rates of childhood asthma events during the 1996 Summer Olympics in Atlanta, Georgia, due to a reduction in local motor vehicle traffic" (Buka, Koranteng, and Osornio-Vargas, 2006). "During the Olympic Games, the number of asthma acute care events decreased 41.6% (4.23 vs. 2.47 daily events) in the Georgia Medicaid claims file," coincident with significant reductions in ozone and other pollutants (Friedman et al., 2001)	"In their primary analyses, which were adjusted for seasonal trends in air pollutant concentrations and health outcomes during the years before and after the Olympic Games, the investigators did not find significant reductions in the number of emergency department visits for respiratory or cardiovascular health outcomes in adults or children." In fact, "relative risk estimates for the longer time series were actually suggestive of increased ED [emergency department] visits during the Olympic Games" (Health Effects Institute, 2010)

Table 6.1 (*Continued*)

Pro (causal interpretation or claim)	Con (counterinterpretation or claim)
"An association between elevated PM10 levels and hospital admissions for pneumonia, pleurisy, bronchitis, and asthma was observed. During months when 24-hour PM10 levels exceeded 150 micrograms/m3, average admissions for children nearly tripled; in adults, the increase in admissions was 44 per cent" (Pope, 1989)	"Respiratory syncytial virus (RSV) activity was the single explanatory factor that consistently accounted for a statistically significant portion of the observed variations of pediatric respiratory hospitalizations. No coherent evidence of residual statistical associations between PM10 levels and hospitalizations was found for any age group or respiratory illness" (Lamm et al., 1994)
"Reductions in respiratory and cardiovascular death rates in Dublin suggest that control of particulate air pollution could substantially diminish daily death....Our findings suggest that control of particulate air pollution in Dublin led to an immediate reduction in cardiovascular and respiratory deaths" (Clancy et al., 2002). "The results could not be more clear, reducing particulate air pollution reduces the number of respiratory and cardiovascular related deaths immediately" (Harvard School of Public Health, 2002)	The same rate of reduction in death rates was already occurring long before the ban and occurred in other parts of Europe and Ireland not affected by it. "Serious epidemics and pronounced trends feign excess mortality previously attributed to heavy black-smoke exposure" (Wittmaack, 2007). "Thus, a causal link between the decline in mortality and the ban of coal sales cannot be established" (Pelucchi et al., 2009)

may not reveal anything about causation, including the correct sign (positive or negative) of the causal influence of exposure on risk, if there is one. For example, if elderly people consume more baby aspirin than younger people to reduce their risk of heart attacks, then level of aspirin consumption might be significantly positively *associated* with increased risk of heart attack, even if increasing aspirin consumption would *cause* reduced heart attack risk at every age.

More generally, causality in risk analysis is not mainly about statistical associations between *levels* of passively observed variables but rather about how *changes*, if they were made, would propagate through systems (Druzdzel and Simon, 1993; Greenland and Brumback, 2002). This distinction should be of critical importance to risk analysts advising public policy decision-makers on the probable consequences of proposed interventions and also to policy makers considering how much weight to give such advice. As a real-world example of how much it matters, mortality rates among the elderly tend to be elevated where and when fine particulate pollutant concentrations are highest among 100 US cities (viz., in cities and months with cold winter days), and yet changes in these pollutant concentration levels from one year to the next are significantly *negatively* associated with corresponding changes in mortality rates, undermining any straightforward causal interpretation of the positive association between pollutant levels and mortality rates (Cox, 2012). Yet, this crucial distinction is often glossed over in the current language and presentation of health risk assessment results. For example, one recent article (Lepeule et al., 2012) announced that for six US cities, "Using the Cox proportional hazards model, statistically significant associations between [fine particulate matter] PM2.5 exposure and all-cause, cardiovascular, and lung-cancer mortality were observed. ...Each 10-μg/m^3 increase in PM2.5 was associated with a 14% increased risk of all-cause death." But the word "increase" here does *not* refer to any actual change (increase over time) in PM2.5 levels or risk over time. Instead, it refers to associations between higher *levels* of PM2.5 and higher levels of risk. The study then infers that "These results suggest that further public policy efforts that reduce fine particulate matter air pollution are likely to have continuing public health benefits." But this causal conclusion about predicted effects of *changes* does not follow from the statistical association between *levels* of PM2.5, since the two may (and in fact, in the United States, often do) have opposite signs (Cox et al., 2013). The contrasting statements on the left and right sides of Table 6.1 suggest that health effects researchers not infrequently leap from observations of associations to conclusion about causation, without carefully checking whether changes in inputs produce the changes in outputs that statistical associations between them suggest. This casual treatment of key causal questions must change, if risk analysis predictions are to become more accurate and trustworthy.

Risk management advice to decision-makers based on past statistical exposure–response associations (or other associations) may not be very useful for correctly predicting probable effects of future changes in exposures (or other variables) brought about by risk management interventions. Instead, an understanding of *causal mechanisms*—that is, of how changes in some variables change others—is usually necessary to correctly predict the effects of interventions (Greenland and Brumback, 2002; Freedman, 2004). This need not be difficult or mysterious. Simulation models (e.g., based on systems of differential and algebraic equations or on discrete-event models such as those in Chapter 3) describing flows of quantities among compartments over time, and the effects of interventions on flow rates, suffice to model the effects of interventions in many practical settings (Druzdzel and Simon 1993; Lu, Druzdzel, and Leong, 2000; Dash and Druzdzel, 2008). However, shifting the emphasis from making judgments about the causal interpretation or "weight of evidence" of statistical associations to rigorous formal testing of causal hypotheses, formulated in terms of propagation of changes along causal paths (or through more complex causal networks), requires a major change in commonly taught epidemiological practices.

HOW TO DO BETTER: MORE OBJECTIVE TESTS FOR CAUSAL IMPACTS

Happily, modern methods of causal analysis now enable decision and risk analysts to address questions about causation by considering relatively objective evidence on how and whether changes in the inputs to a system propagate to cause changes in its outputs. This is a far more useful, and objective, approach than making judgments about statistical associations, for reasons given next. Well-developed methodological principles for drawing sound causal inferences from observational data include asking (and using data to answer) the following simple, systematically skeptical questions about observed exposure–response associations to test whether the observations are logically capable of providing evidence for a genuine causal relation:

- *Do the study design and data collected permit convincing refutation of noncausal explanations for observed associations* between levels of exposure and response (or between levels of other hypothesized cause-and-effect variables)? Potential noncausal explanations for associations include data selection and model selection biases, residual confounding by modeled confounders, unmodeled confounders, unmodeled errors in exposure estimates and covariates, unmodeled uncertainties in model form specification, regression to the mean, and so forth (Cox, 2007). These potential rival explanations can be ruled out by appropriate study designs, control group comparisons, and data analyses, if they indeed do not explain the observed associations (Campbell and Stanley, 1966; Maclure, 1990; Cox, 2007). Assuming that they have been ruled out, the next questions consider whether there is objective evidence that the observed relation might be causal.

- *Are significant positive associations also found between changes in exposures and changes in response rates?* If the answer is no, as revealed in some panel data studies of previously reported positive associations between exposure and response levels (Stebbings, 1978), then this undermines causal interpretation of the positive associations.

- *Do changes in hypothesized causes precede changes in their hypothesized effects?* If not, for example, if health effects are already declining before reductions in exposure, then this casts doubt on the latter being a cause of the former. Doubt is increased if, as in the Dublin study in Table 6.1, a steep reduction in exposure is not followed by any detectable corresponding change in the rate of decline in effects.

- *Are reductions in hypothesized effects significantly greater in times and places where exposure went down than where exposure remained the same or went up?* If not, as in the HEI (2010) analysis of the Atlanta Olympics data in Table 6.1, then this casts doubt on the hypothesis that reductions in exposure caused the reductions in effects.

- *Do changes in hypothesized causes (e.g., exposures) help to predict subsequent changes in their hypothesized effects?* If not, for example, if changes in effects appear to be statistically independent of previous changes in the hypothesized causes, then this reduces the plausibility of a causal interpretation for a regression model, or other statistical model, relating them.

Such qualitative questions provide clear commonsense and logical foundations as screens for causal inference, and they are relatively easy to understand and ask.

Quantitative methods, although sometimes technically sophisticated, help to implement many of the same basic ideas and to provide relatively objective answers using formal statistical tests. Among the most useful analytic methods for testing causal hypotheses and constructing valid causal models from data are the following:

- *Intervention analysis* (Friede, Henderson, and Kao, 2006), also called interrupted time series analysis, tests whether the best-fitting model of the data-generating process for an observed time series, such as daily mortality and morbidity counts, changed significantly at, or following, the time of an intervention. Intervention analysis provides methods to identify, test for, and estimate significant changes in time series that might have been caused by an intervention and that cannot easily be explained by other (noncausal) hypotheses.

- *Change-point analysis* (Helfenstein, 1991; Gilmour et al., 2006). Intervention analysis seeks to quantify the changes in effects following known changes in hypothesized causes. By contrast, change-point analysis searches for any significant changes in the data-generating process for effects (e.g., mortality rates) over an interval of time—for example, a change in the slope of a long-term declining trend in cardiovascular mortality rates or a change in the season-specific rate of hospital admissions for pediatric asthma. If such a change point is detected at, or closely following, an intervention, such as an emissions ban that reduces pollution levels, then the intervention might have caused the change. If no change is detected, then there is no evidence that the intervention had a detectable effect.

- *Quasiexperimental designs* and methods (Campbell and Stanley, 1966) make use of control group comparisons (including pre- and postintervention observations on the same subjects) to try to systematically refute each of a list of identified methodological threats to valid causal inference, such as "history" (e.g., that the Dublin coal ban occurred during a long-term historical trend toward lower cardiovascular mortality rates due to better prevention, diagnosis, and treatment), regression to the mean (unusual bursts of ill effects tend to be followed by lower levels even if any intervention that they may have triggered have no effect), aging of subjects, and so forth.

- *Panel data analysis* (Angrist and Pischke, 2009) examines how well *changes* in explanatory variables predict *changes* in responses using repeated measures of the same observational units over time to control for unobserved confounders. In health risk assessment, comparing changes in exposures to changes in responses can give a very different understanding of the likely health consequences caused by changes in exposure than studying estimated (or assumed) statistical associations between exposure and response levels (Stebbings, 1978; Cox et al., 2013).

• *Granger causality tests* (Eichler and Didelez, 2010). Changes in causes should help to predict subsequent changes in their effects, even if there is no intervention in the time series being observed. To formally test whether changes in exposure might be a contributing cause of changes in short-term daily mortality rates, for example, one could compare a simple predictive model, created by regressing future mortality rates against their own past (lagged) values, to a richer model that also regresses them against lagged values of exposure as possible predictors. If including exposure history does not improve predictions of mortality rates (e.g., producing smaller mean squared prediction errors or larger mutual information between predicted and observed values), then the time series data do not support the hypothesis that exposure causes mortality, in the sense of helping to predict it. This method of testing causal hypotheses is incorporated in the *Granger causality test*. (It is now widely and freely available, e.g., as the *granger.test* procedure in R.) In practice, Granger causality testing may show that some significant correlates of short-term mortality rates (such as low temperature (Mercer, 2003)) are also Granger causes of the short-term mortality rates, while others (e.g., PM2.5) are not Granger causes (Cox et al., 2013). Although Granger tests are subject to the usual limitations of parametric modeling assumptions, such as the use of a linear regression model, the lack of Granger causation between exposure and response even when there is a clear, statistically significant positive regression relation between them highlights the importance of distinguishing between positive regression relations and causal relations. (This distinction has not been prominent in the air pollution health effects accountability literature to date, but deserves to be in the future.)

• *Conditional independence tests* (Friedman and Goldszmidt, 1998; Freedman, 2004). In both cross-sectional and longitudinal data, a cause should provide some information about its effect that cannot be obtained from other sources. Conversely, if an effect is *conditionally independent* of a hypothesized cause, given the values of other explanatory variables (e.g., measured potential confounders and covariates), then the causal hypothesis is not supported by the data (Freedman, 2004). For example, if daily mortality rates are conditionally independent of pollution levels, given city and month and temperature, then there would be no evidence that pollution levels make a causal contribution to daily mortality rates. Conversely, if there is no way to eliminate the significant difference between mortality rates for very different pollutant levels, holding other covariate levels fixed, then pollutant levels would appear as direct causes ("parents") of daily mortality rates in causal graph models (Friedman and Goldszmidt, 1998; Freedman, 2004).

- *Counterfactual and potential outcome models.* A possible measure of causal impact of exposure on mortality rates in a population is the difference between the average mortality rates with and without exposures. Since no individual can be both exposed and unexposed, estimating this difference typically requires considering counterfactual exposures and responses. One way to create such estimates is with a regression model. More generally, regression models can be used to estimate how risk would have changed had exposures been reduced, provided that the regression models describe effects of counterfactual exposures, rather than merely associations between predictors and predicted variables. There has much recent progress in technical methods for developing and fitting such counterfactual regression models ("marginal structural models" and their extensions) to data to predict what would have happened if exposure had been lower or absent (Robins, Hernán, and Brumback, 2000; Moore et al., 2012). Such counterfactual causal models can yield insights and conclusions that are quite different from earlier regression models. For example, in one recent study, adverse effects of ozone exposure that are statistically significant in earlier regression models (which must make unverifiable modeling assumptions about what responses would be to combinations of predictors that do not occur in reality) are not significant when methods are applied that only use realistic exposure–response data (Moore et al., 2012).

- *Modeling causal mechanisms via propagation of changes through causal chains or networks.* If exposure causes adverse health effects, it must do so via one or more causal pathways. Collecting biomarker data can allow testing of specific causal hypotheses about the mechanisms of harm (Hack et al., 2010). Causal graph models, which factor joint distributions into marginal and conditional distributions (Freedman, 2004), can be constructed to preserve causal orderings from structural equations or mathematical mechanistic models (Druzdzel and Simon, 1993; Lu, Druzdzel, and Leong, 2000; Dash and Druzdzel, 2008). Then, if predicted changes in the variables that are supposed to transmit causal impacts are not observed, this would provide evidence against the hypothesized causal mechanism. Conversely, detecting and quantifying those changes (via conditional probability tables (CPTs)) allows prediction of the sizes of changes in health effects to be expected from changes in exposure, given the values of other variables in the causal model. For example, a recent study (Chuang et al., 2007) provided panel data to test the specific mechanistic hypotheses for air pollution health effects, including that PM2.5 causes adverse cardiovascular effects by increasing oxidative stress as measured by urinary 8-hydroxy-2′-deoxyguanosine (a marker for oxidative DNA damage). As summarized by Kaufman (2007), "blood and electrocardiographic markers were repeatedly collected over 3 months to examine multiple potential mechanistic pathways. They had the benefit of fairly large daily fluctuations in exposure, presumably dictated by meteorological conditions. While their inflammatory, oxidative stress, fibrinolysis, and coagulation health markers did not change consistently as hypothesized with fine particles, they did detect associations with some PM components and credited these to traffic-related air pollution.... Their measure of "oxidative stress" (urinary 8-hydroxy-2′-deoxyguanosine, assessing oxidative DNA damage) was not associated with pollution exposures. Heart rate variability metrics, on the other hand, consistently demonstrated negative associations with all air pollutants examined, in a manner that appeared to be independent of inflammation." This ability to refute expected causal hypotheses, to discover new potential threats (such as reductions in heart rate variability), and to reveal unexpected time-ordered sequences of changes makes panel data especially valuable for learning from data by testing and improving mechanistic models.

Table 6.2 summarizes some of the best-developed quantitative methods for testing causal hypotheses and for quantifying the sizes of causal effects. These methods of causal analysis are relatively objective. Unlike expert judgments and opinions about the causal interpretation of statistical associations, they can be independently replicated by others

Table 6.2 Some formal methods for modeling and testing causal hypotheses

Method and references	Basic idea	Appropriate study design
Conditional independence tests (Friedman and Goldszmidt, 1998; Freedman, 2004)	Is hypothesized effect statistically independent of other ("ancestor") variables, given values of hypothesized direct causes ("parents") in causal graph model? If so, this strengthens causal interpretation Example: Is dependence of disease risk on location explained by variations in measured exposures at different locations? Is hypothesized effect statistically independent of hypothesized cause, given values of other variables? If so, this undermines causal interpretation Example: Does effect of exposure on disease risk disappear after accounting for education and income?	Cross-sectional data Can also be applied to multiperiod data (e.g., in dynamic Bayesian networks)
Panel data analysis (Stebbings, 1978; Angrist and Pischke, 2009)	Are changes in exposures followed by changes in the effects that they are hypothesized to help cause? If not, this undermines causal interpretation; if so, this strengthens causal interpretation Example: Are changes in air pollutant levels in different cities followed (but not preceded) by corresponding changes in respiratory mortality rates?	Panel data study: Collect a sequence of observations on same subjects or units over time

(continued)

Table 6.2 (*Continued*)

Method and references	Basic idea	Appropriate study design
Granger causality test (Eichler and Didelez, 2010)	Does the history of the hypothesized cause improve ability to predict the future of the hypothesized effect? If so, this strengthens causal interpretation; otherwise, it undermines causal interpretation. Example: Can daily mortality rates in different cities be predicted better from time series histories of daily pollutant levels and mortality rates than from the time series history of daily mortality rates alone?	Time series data on hypothesized causes and effects
Quasiexperimental design and analysis (Campbell and Stanley, 1966)	Can control groups and other comparisons refute alternative (noncausal) explanations for observed associations between hypothesized causes and effects? For example, can coincident trends and regression to the mean be refuted as possible explanations? If so, this strengthens causal interpretation	Longitudinal observational data on subjects exposed and not exposed to interventions that change the hypothesized cause(s) of effects
Intervention analysis, change-point analysis (Helfenstein, 1991; Gilmour et al., 2006)	Does the best-fitting model of the observed data change significantly at or following the time of an intervention? If so, this strengthens causal interpretation	Time series observations on hypothesized effects and knowledge of timing of intervention(s)

	Do the quantitative changes in hypothesized causes predict and explain the subsequently observed quantitative changes in hypothesized effects? If so, this strengthens causal interpretation Example: Do cardiovascular disease mortality rates fall significantly faster in cities that banned smoking in public areas than in cities that did not, with the rate of decrease jumping after the ban in each case?	Quantitative time series data for hypothesized causes and effects
Counterfactual and potential outcome models (Robins, Hernán, and Brumback, 2000; Moore et al., 2012)	Do exposed individuals have significantly different response probabilities than they would have had if they had not been exposed? Example: Do inhabitants of a city that bans coal-burning have lower mortality risk after the ban than they would have had without it?	Cross-sectional and/or longitudinal data, with selection biases and feedback among variables allowed
Causal network models of change propagation (Dash and Druzdzel, 2008; Hack et al., 2010)	Do changes in exposures (or other causes) create a cascade of changes through a network of causal mechanisms (represented by equations), resulting in changes in the effect variables? Example: Do relatively large variations in daily levels of fine particulate matter (PM2.5) air pollution create corresponding variations in markers of oxidative stress in the lungs?	Observations of variables in a dynamic system out of equilibrium

(continued)

Table 6.2 (Continued)

Method and references	Basic idea	Appropriate study design
Negative controls (for exposures or for effects) (Lipsitch, Tchetgen Tchetgen, and Cohen, 2010)	Do exposures predict health effects better than they predict effects that cannot be caused by exposures? Example: Does being vaccinated against influenza have a stronger association with reduction in influenza hospitalizations than with reduction in trauma hospitalizations? Are hospitalizations reduced more during flu season than before or after flu season? If not, then the association may reflect biases or confounding (e.g., perhaps relatively low-risk patients are more likely to be vaccinated)	Observational studies

using standard statistical methods (such as *granger.test* in the R statistical computing environment). They focus on answering the following key factual questions:

1. *Can any effect be detected (e.g., a significant change in a health effects time series following a change in exposures)?* Change-point analysis, intervention analysis, and panel data analysis address this question. If there is no apparent effect, as in the Dublin study data (Wittmaack, 2007), then there is nothing to explain, and proffered causal interpretations are superfluous.

2. *If so, how large is it?* This may be assessed via intervention analysis, change-point models, panel data, or quasiexperimental prepost comparisons. Counterfactual causal models can be used to estimate effects specifically caused by exposure, after untangling effects of confounders and feedback loops (Moore et al., 2012). If this causal effect is only a fraction as large as the statistical "effect" estimated from a regression model, then only that fraction of the statistical association should be attributed to exposure, as opposed to confounding or other sources.

3. *Can changes in responses be explained or predicted just as well without knowledge of a putative cause as with it?* This crucial screening question can be answered using Granger tests, conditional independence tests, and quasiexperimental analyses. If knowledge of changes in a hypothesized cause does not improve ability to predict its hypothesized effects or, conversely, if the observed effects can be fully explained by other variables, then the causal hypothesis is not supported.

4. *Are the sequences of changes in variables implied by a hypothesized causal mechanism observed?* This can be addressed using causal graph models and panel data analysis applied to biomarker data.

Using modern methods of causal analysis to address these factual questions can liberate risk analysts and policy makers from the need to rely on (potentially biased or unreliable) subjective judgments in addressing questions of causality. (See Chapters 2 and 10 for further discussion of potential pitfalls of subjective judgments.) They provide an alternative to the traditional Hill-type criteria popular in epidemiology (such as strength, consistency, specificity, and temporality of associations).

PREDICTIVE MODELS: BAYESIAN NETWORK (BN) AND CAUSAL GRAPH MODELS

A useful innovation for extending the realism and flexibility of graphical causal models moves beyond expected values and assumed linear regression relations among variables (used for much of the twentieth century in statistical and econometric analyses of causation via path analysis and structural equations models), by instead letting variables represent arbitrary random variables, with the *conditional probability distribution* of each variable depending (only) on the values of the variables that point into it. This is the essential idea exploited in *BN* models. Each node in a discrete BN model can be thought of as having a corresponding *conditional probability table* (CPT) giving the conditional probability of each of its possible values for each combination of its input values. (In practice, if the node value probabilities are not distinct for all possible combinations of inputs, then more efficient data structures than explicit CPTs may be used, but such implementation details are hidden in the software that makes BN technology convenient for general users. For example, if data are sufficiently abundant, then the empirical CPTs for nodes may be stored as classification trees and estimated by classification tree algorithms (Meek and Thiesson, 2010), which can also be used to check the local conditional independence relations implied by the BN (Friedman and Goldszmidt, 1998).)

Changing the value (or the probability distribution of values) for one or more of the input variables in a BN causes the distributions for the values at other nodes to change in response to changes in the marginal distributions of their inputs. (Thus, in the two-node BN $X \rightarrow Y$, the marginal probability that $Y = y$, denoted by $\Pr(y)$, is $\Pr(y) = \sum_x \Pr(y \mid x)\Pr(x)$, so changes in the marginal distribution of input values, $\Pr(x)$, change $\Pr(y)$ accordingly.) Such changes can propagate along directed paths throughout the network, giving new probability distributions for output values and intermediate nodes as the input values (or, if they are uncertain, their probability distributions) change.

Example: A BN Model for Benzene Exposure–Response Relations

Figure 6.1 shows a simplified summary of the structure of a BN model for the causal relation between inhalation exposure to benzene and changes in the probable values of several other variables, including an

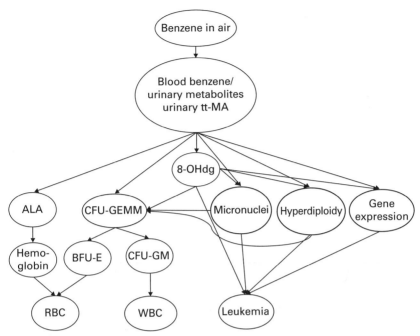

Figure 6.1   An Example Bayesian network (BN) causal model for benzene and leukemia risk. *Source:* Hack et al. (2010).

indicator of whether the exposed individual develops leukemia. The graphical model is simplified by lumping together into one node several intermediate variables (such as multiple urinary metabolites, in addition to trans, trans-muconaldehyde (ttMA), or multiple indicators of gene expression profiles) and by suppressing the detailed time courses of all variables (which could be represented by a dynamic Bayesian network (DBN) model, discussed shortly). The *structure* of the BN shows that the probability of leukemia depends on gene expression, chromosomal aberrations (hyperdiploidy), micronuclei, and the level of oxidative stress (as indicated by 8-oxo-2′-deoxyguanosine (8-OHdg), a product of DNA oxidation). Changes in the exogenous input, "Benzene in Air," cause the conditional probability distributions of these other variables to change, which in turn changes the probability of leukemia.

Changes in benzene exposure also change the conditional proba-bility distributions of other variables, such as those for the sizes of the populations of the erythroid burst-forming units (BFU-E) and the gran-ulocyte–macrophage colony-forming units (CFU-GM) in the bone marrow or the size of the circulating population of red blood cells (RBC) and white blood cells (WBC). Although these variables do not lie on the directed paths leading from benzene exposure to leukemia, they share common ancestors (such as 8-OHdg and hyperdiploidy) with the leukemia node, and thus, one might expect strong statistical associations between these biomarkers of benzene exposure and risk of leukemia, even though neither is shown as a potential cause (ancestor in the *directed acyclic graph* (DAG) model) of the other.

BNs are usually assumed to be acyclic (DAGs), which makes the forward propagation of changes in their input values or distributions especially straightforward. For much practical work, Gibbs sampling or other Markov chain Monte Carlo (MCMC) methods (implemented in software such as WinBUGS or R) provide a fast, practical way to approximate, as accurately as desired, the output distributions for any given set of input values (or joint distribution of uncertain input values). When feedback loops or other cycles are required to adequately model a system, the successive values of each variable in consecutive periods or "time slices" can be represented as distinct variables. The BN relating these time-stamped variable values is then acyclic. This construction is called a *dynamic Bayesian Network* (DBN), and it allows BN technology to be applied to realistically complex dynamic systems. (The underlying algorithms for quantifying conditional probabilities of outputs given inputs for DBNs extend those used for static BNs. For example, instead of relying on Gibbs sampling, as in WinBUGS, the use of efficient importance-weighted sampling methods such as *dynamic particle filtering*, developed specifically for multiperiod models with consecutive conditioning of successive variable values on period-specific events or information, may be used to improve the speed and accuracy of dynamically estimating posterior distributions. In addition to discrete nodes, with their CPTs, there may also be continuous nodes, with regression models in place of CPTs, in which the continuous output value of a variable is usually assumed to be conditionally Gaussian, with mean and variance that may depend on the values of the parent nodes that point into it. Das et al. (2005) discuss tractable inference algorithms, using particle filtering and other techniques, for hybrid DBNs with both discrete and continuous variables, and apply them to real-time updating of a threat assessment for possible ambush by mobile attackers in an urban environment.)

Both exact and approximate inference algorithms for BNs and DBNs are available that let the distributions of any variable(s) in the model be updated (via Bayesian inference, i.e., conditioning in accord with Bayes' rule) based on any set of observed (or assumed) values for other nodes (e.g., Cheng and Druzdzel, 2000; Guo and Hsu, 2002; Liu and Soetjipto, 2004). Such practical Bayesian inference in BNs directly addresses a key technical challenge of probabilistic risk assessment (PRA): how to calculate the probability distributions for unobserved quantities (e.g., future disease states) from observed or assumed values for other quantities (e.g., current exposure metrics and biomarker and other covariate data) using knowledge of the conditional probabilistic relations (represented in the CPTs) and the structure of causal pathways or dependency relations (represented via the BN graph structure) relating them. For example, in the benzene model in Figure 6.1, a risk assessment that quantifies the conditional probability of developing leukemia, given an exposure to benzene in air, could also be informed by (i.e., conditioned on) the results of blood tests that provide information on RBC and WBC, even though these are not themselves causal prede-cessors of leukemia. Such Bayesian inference, in which posterior probabilities for events or conditions of interest (such as disease occur-rence) are conditioned on all available measurements of variables in a BN, can be performed routinely using these BN algorithms.

Inference about the probability distributions of values of some nodes (i.e., variables), given the values of others, is the essence of BN infer-ence when a BN model is known (or assumed). Although it is often natural and tempting to interpret BN graph structures as showing the directions of causal influences among variables, the technology of Bayesian inference in BNs works equally well with or without such a causal interpretation: the fundamental inference mechanism is condi-tioning of the probabilities of some variables on observed values of others, and this need not reflect causality, in the sense that changes in parent nodes need not bring about (or make more probable the occur-rence of) changes in their children. Such a causal relation may lend itself naturally to representation via a directed arc in a BN, but BN calcula-tions can be carried out even when a causal interpretation does not hold. In this case, the BN is simply an efficient way of factoring and making calculations with a joint probability density function for the variables.

A more challenging type of inference is to learn the BN model itself—its structure (the graph showing which nodes point into which others), as well as the CPTs at its nodes—from data. There are now many BN-learning algorithms. Because the CPTs impose no restrictions (such as linear relations between expected values) on the conditional distributions of outputs given inputs, BNs provide a very flexible, nonparametric framework for modeling probabilistic relations between discrete random variables. This flexibility supports methods for learning BNs from data by testing whether the empirical conditional frequency distribution of a variable differs significantly for different combinations of levels of other variables. If so, then the relevant conditioning variables are identified as potential direct causes (parents), direct consequences (children), or indirect causes (ancestors) or consequences (descendants) of the affected variable; otherwise, no arrows or paths need to connect them. To further identify which variables points into which, starting from the statistical finding that two variables provide significant mutual information about another (meaning that the conditional distribution of one depends on the value of the other), a variety of algorithms have been developed for learning BNS from data, typically by maximizing a statistical score subject to some simplifying assumptions or constraints (e.g., Daly, Shen, and Aitken, 2011). The richness of the options that are now available is well illustrated by the following description of *bnlearn*, an R package for BN learning and inference (http://www.bnlearn.com/):

> *bnlearn* implements the following *constraint-based structure learning algorithms*: Grow-Shrink (GS); Incremental Association Markov Blanket (IAMB); Fast Incremental Association (Fast-IAMB); Interleaved Incremental Association (Inter-IAMB); the following *score-based structure learning algorithms*: Hill Climbing (HC); Tabu Search (Tabu); the following *hybrid structure learning algorithms*: Max-Min Hill Climbing (MMHC); General 2-Phase Restricted Maximization (RSMAX2); the following *local discovery algorithms*: Chow-Liu; ARACNE; Max-Min Parents & Children (MMPC); and the following *Bayesian network classifiers*: naive Bayes. Discrete (multinomial) and continuous (multivariate normal) data sets are supported, both for structure and parameter learning. …Each *constraint-based algorithm* can be used with several conditional independence tests [e.g., mutual information, Chi square, and Akaike Information Criterion test]… and each *score-based algorithm* can be used with several score functions [including] loglikelihood, the Akaike Information Criterion (AIC), the Bayesian Information Criterion (BIC), [and] a score equivalent… posterior density (BDe).

Thus, current BN-learning software offers several substantial technical solutions to the challenges of learning BN structures from data, and design trade-offs for BN-learning algorithms (e.g., between the computational ease of fitting maximum-likelihood models and the greater difficulty of finding models that reduce prediction error rates) have been studied in a large and still expanding literature. Even when data sets are insufficient to confidently identify a single best BN model (by some criterion), recent *model ensemble* methods, including Bayesian model-averaging methods for BNs, allow final predictions to be based on multiple plausible BNs, which can reduce prediction errors, compared to selecting any single model (Liu, Tian, and Zhu, 2007; Daly, Shen, and Aitken, 2011).

In practice, incorporating knowledge-based constraints (e.g., a causal ordering or partial ordering of variables, such as that benzene inhalation might cause changes in WBC counts, but changes in WBC counts do not cause benzene inhalation) can greatly reduce the computational search effort needed to identify one or more BNs that describe the available data, significantly improving both the accuracy (as judged by statistical criteria) and the causal interpretability of the resulting BNs (Daly, Shen, and Aitken, 2011). Such knowledge might come from an underlying mathematical model of the causal process, for example, using the Causal Ordering Algorithm (COA) (Dash and Druzdzel, 2008) or from commonsense understanding of potential causes and consequences. When correct causal information is not known, however, even diligent and ingenious efforts to apply BN and DBN technology are sometimes disappointing. Perhaps most notably, concentrated and sustained efforts by many experts to use BNs and DBNs to automatically infer (or "reverse engineer") descriptions of gene interaction networks from observed time series of gene activation or expression have as yet met with only very limited success (David and Wiggins, 2007).

The maturation of BN technology has led to many successful applications in PRA and decision support, however. Some important examples include:

- *Disease diagnosis and management.* BN-based risk assessment and risk management advisory systems have been developed for prostate cancer (Regnier-Coudert et al., 2012), lung cancer (Jayasurya et al., 2010), breast cancer (Burnside et al., 2009), and brain tumors (Weidl, Iglesias-Rozas, and Roehrl, 2007), among others.
- *Natural disaster and catastrophe risk analysis.* Applications include avalanche risk assessment systems (Gret-Regamey and Straub, 2006);

better flood prediction, with reduced false alarms and failures to warn compared to other methods (Li et al., 2010); and seismic risk prediction, integrating a variety of different data and knowledge sources (including mining of spatial data) (Cockburn and Tesfamariam, 2012; Li et al., 2012).

- *Process control and safety management* for industrial processes, from hazardous chemical or manufacturing facilities to food safety (Albert et al., 2008; Smid et al., 2012).

- *Transportation accident and maritime risk analysis*, for example, modeling effects of organizational factors and other risk factors on train accidents (Marsh and Bearfield, 2004) or assessing collision and accident risks for offshore platforms and vessels (Ren et al., 2008).

- *Risk assessment of software quality, reliability, and defects*, by using a BN to integrate judgments about such hard-to-quantify (but easy-to-rate) inputs as testing process quality, testing effort, and quality of documentation (Fenton, Neil, and Marquez, 2008).

- *Financial and credit risk analysis and bank stress testing*, in which BNs help to identify business interdependencies (and hence potential opportunities for risk contagion or common mode failures) among groups of borrowers, thus helping banks to avoid the inadvertent concentration of exposures to positively correlated credit risks in a bank's loan portfolio (Pavlenko and Chernyak, 2010).

- *Organizing expert impressions* about a variety of traditionally hard-to-quantify business, financial, engineering, operational, and adversarial risks, such as supply chain and supplier risks (Lockamy, 2011), new product development risks (Chin et al., 2009), cybersecurity, and terrorism attacks (Ezell et al., 2011).

Despite many successful decision and risk management applications of BNs, BN technology can readily be misused by applying it to poorly defined or meaningless concepts, for which human experts may nonetheless comfortably provide subjective estimates of conditional probabilities and postulate causal relationships. For example, nothing would prevent an analyst from soliciting opinions on the "degree of harshness" of a company's business environment, the "level of preparation" that the business has to meet its challenges, and the resulting conditional probabilities of different levels of "business success," given opinions about "degree of harshness" and "level of preparation." These concepts might have no clear definitions or meanings or might mean mutually inconsistent things to different experts and users, and yet a BN model could still be built and used and perhaps even foster a reassuring illusion of meaningful risk analysis, if no one pressed for meaning or clarity or realized that the inputs and outputs lacked useful definition. The ability of BNs to represent, store, process, and produce nonsense in the guise of elegantly displayed probability distributions invites the creation of poorly thought-out applications of BNs, especially since even well-developed applications often combine elements of expert judgment with more objective data or statistical models to assess conditional probabilities and to structure the BNs.

DECIDING WHAT TO DO: INFLUENCE DIAGRAMS (IDS)

BN risk models predict probabilities of outputs given inputs and/or other observations. To support improved risk management decisions, *influence diagrams* (IDs) (Howard and Matheson, 2005) extend BNs to include *choice nodes* (where decisions are made) and *value nodes* (where consequences are evaluated), as well as BN-type *chance nodes* (representing random variables whose conditional probability distributions depend on the values of the parent nodes pointing into them). ID software products, such as the free Structural Modeling, Inference, and Learning Engine (SMILE) and GeNIe development environment for graphical interfaces (http://genie.sis.pitt.edu/) or the commercial product *Analytica* (http://www.lumina.com/), also allow deterministic formulas (i.e., *deterministic nodes*), which may be viewed as special cases of chance nodes that assign a probability mass of 1 to the output value determined by the input values. Algorithmically, an ID can be transformed to an equivalent BN by suitable recoding of choice and value nodes as chance nodes, and well-developed algorithms for inference in BNs can then be applied to infer optimal decisions (Zhang, 1998).

IDs incorporate the flexible modeling capabilities of BNs for representing probabilistic causal mechanisms as CPTs (as well as the capacity to represent noncausal probabilistic dependencies of observed values, induced by a statistical joint distribution of values for variables that need not all be mutually statistically independent). They support the same types of Bayesian inference as BNs (i.e., conditioning on information to update posterior probability distributions of model variables). But ID algorithms and software also allow automatic optimization of *decisions* (i.e., recommended choices at the choice nodes to maximize expected utility). The importance of causation is especially clear in this context: IDs seek to identify and recommend choices that will cause preferred probability distributions for consequences. Several workers in artificial intelligence, most notably Pearl (2010) and coworkers, have remarked that most traditional statistical and epidemiological methods lack adequate concepts and language for predicting probable consequences in situations where a decision-maker chooses to *do* something or *set* a value of a control variable, rather than merely observing values of variables; they have therefore proposed new notation (such as $P(y \mid do(x))$ for the probability that variable Y will take value y if the decision-maker sets the value of X to x) and causal graph methods to model the probabilities of effects caused by interventions. These are precisely the probabilities needed to populate the nodes of an ID model in which a choice affects the probability distribution of one or more other variables.

An ID can be used not only to show the qualitative paths by which decisions and other causes affect consequences (via the graph structure of the ID DAG model) but also to quantify the probabilities and expected utilities of consequences from different decisions. This combination of qualitative and quantitative information provides an explicit rationale for recommended decisions. Most ID software products also provide facilities for sensitivity analyses to show how the optimal decision changes as input assumptions, data, or values (e.g., trade-off weights at value or utility nodes) change. If the decisions of multiple agents—for example, a terrorist attacker and a defender, a physician and a patient (and possibly other parties, such as a drug manufacturer and an insurance company or ACO), a corporation and a regulator, or a smuggler and an inspector—cause the consequences of interest, then IDs can be extended to represent the beliefs, preferences (utilities), and decisions of multiple agents. Such *Multiagent Influence Diagrams* (MAIDs) represent the beliefs and preferences of many interacting decision-makers, allowing game-theoretic and/or descriptive modeling of how the choices of one decision-maker affect the choices of others and hence the probable consequences of the multiagent interaction (Koller and Milch, 2003; Gal and Pfeffer, 2008).

As discussed in Chapter 3, when enough is known about a system to simulate its behavior in detail, for example, using discrete-event simulation modeling, then probabilistic relations between inputs and outputs similar to those that an ID provides can be developed via stochastic simulation, so that the response surface or probabilistic input–output relation *Pr(outputs | inputs)* can be quantified numerically. Recent techniques of simulation–optimization can then be used to identify the input combinations that maximize expected utility (see Chapter 3). Thus, an ID can be viewed as similar to a high-level simulation–optimization model, except that its CPTs need not be populated via detailed simulation or knowledge of causal mechanisms, but may instead be derived from empirical data or may be guessed at by users if adequate data are lacking.

IDs have been applied for over two decades in medical decision-making and disease risk management expert systems (Neapolitan, 1991; Owens and Nease, 1993; Owens, Shachter, and Nease, 1997). They have been applied more recently to public health risk management decisions, such as whether, when, and how to vaccinate elderly populations against influenza (e.g., Baio et al., 2006) or manage polio risks (Tebbens et al., 2008). Recent applications in ecological risk analysis include analysis of Deepwater Horizon spill responses (Carriger and Barron, 2011) and deciding how to minimize adverse effects of pesticides while meeting other goals (Carriger and Newman, 2012). Since they link decisions directly to their probable consequences and to evaluations of the resulting consequence probabilities, IDs and simulation–optimization causal models provide ideal support for many risk analysis applications; it is thus perhaps surprising that they are not yet even more widely in PRA and risk management.

WHEN IS A BN OR ID CAUSAL?

Like BNs, IDs are vulnerable to abuse. It is easy to incorporate mistaken or unvalidated modeling assumptions, treat noncausal (e.g., reduced-form statistical or econometric regression) relations as if they were causal, or show plausible-looking pictures that are vague about exactly whom or what they describe. For example, whereas a discrete-event simulation model is usually very clear about the distinction between *variability* (arising from the joint distribution of attribute values for the individual entities in the model) and *uncertainty* (arising from stochastic events and from the uncertainty distributions of input values), IDs and BNs may blur these important conceptual distinctions.

In the domain of public health, Greenland and Brumback (2002) note that causal graphs (which include BNs and IDs) are often applied to populations, rather than to individuals within populations, or to causal mechanisms within individuals. An ID based on empirical relations in populations may not correctly describe causal relations for any individuals in the population, as in the baby aspirin heart attack risk example.

However, these limitations are largely avoidable, at least in principle. Causal graph models, including BNs and IDs, can be constructed to represent causal mechanisms via their graph structures and CPTs (Druzdzel and Simon, 1993; Lu, Druzdzel, and Leong, 2000). (Technically, every BN can be represented by an equivalent system of simultaneous structural equations, by recoding its CPTs as structural equation model (SEM) equations; a COA for SEMs (Dash and Druzdzel, 2008) can then be applied to this SEM. The causal graph structure of the variables in the SEM, showing which variables are determined by which others, is the same as the causal graph structure of the BN. The equations of the SEM represent causal mechanisms, in the usual SEM sense that changes in their right-hand-side variables create corresponding changes in the left-hand-side variables to restore equality. If the BN graph structure is constrained to correspond to valid structural equations, that is, equations that correctly model causal mechanisms, then the BN may be interpreted causally (Druzdzel and Simon, 1993)). Thus, although general BNs simply represent ways to factor joint probability distributions, so that their arrows do not necessarily have any causal interpretation (since arc directions can be reversed, via application of Bayes' rule, without changing the joint distribution being represented), both BNs and IDs (Lu, Druzdzel, and Leong, 2000) can be deliberately constructed to represent networks of causal mechanisms.

However, making sure that an ID has such a valid causal interpretation takes care. One essential requirement is that the directions of arrows and the contents of CPTs should correctly reflect directed causal mechanisms (so that changes in inputs to a node really do cause changes in the probability distribution of its values) rather than just statistical associations or probabilistic conditioning. This can be accomplished, for example, by first constructing a more detailed simulation model of the underlying causal processes and then constructing an ID that faithfully preserves the causal ordering of the variables while simplifying and summarizing the relations among changes in node inputs and probable values using CPTs. A second requirement is that the detailed CPTs or regression relations used to populate ID nodes should represent causal effects of inputs on probable outputs rather than simply statistical associations. For example, if a regression model is used to estimate how a node's value depends on the values of its parents, then it should be a structural equation rather than a reduced-form equation. Otherwise, changes in inputs may fail to cause the changes in output probabilities predicted by the model.

Example: An ID for Health Effects of Air Pollution

Figure 6.2 shows an illustrative ID model for evaluating air pollution control health effects and optimizing emissions reduction decisions (Mansfield, Sinha, and Henrion, 2009). In this model, emissions reduction choices (decision rectangle, upper left corner) affect the net present value (NPV) of net benefits (hexagonal value node, right side) via increases in control costs and reductions in mortality. Figure 6.3 (Morgan and Henrion (1990, chapter 10)) plots probability bands for the distribution of total costs, rather than net benefits (based on monetized values of adverse health effects and control costs, in a similar ID model), against the decision variable indicating the fraction of

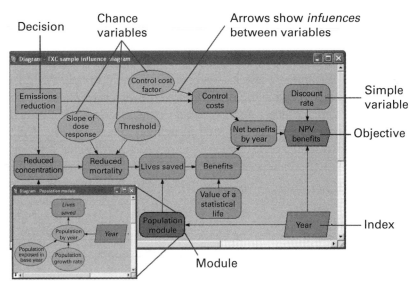

Figure 6.2   An illustrative ID for net benefits of air pollution control. *Source:* Mansfield, Sinha, and Henrion (2009).

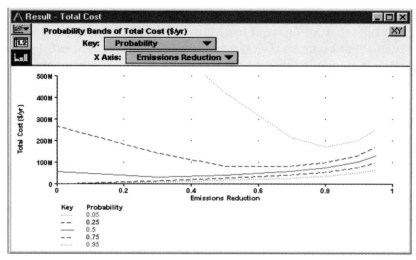

Figure 6.3   ID output identifies the emissions reduction that minimizes expected net cost. *Source:* Morgan and Henrion (1990).

emissions eliminated. (For this hypothetical example, the median of the total cost distribution is minimized for an emissions reduction factor of about 0.3.) Such information can potentially help decision-makers to identify pollution reduction targets that will balance marginal costs and expected marginal benefits while accounting for various uncertainties.

ID risk and uncertainty analysis software clearly has great potential to facilitate probabilistic risk assessment. It can help to help visualize and communicate probable consequences of alternative choices (and the extent of uncertainty about probable consequences). It can also identify choices that maximize criteria such as expected utility or that minimize criteria such as expected total costs. Yet, the same software can create models that incorporate unvalidated or mistaken assumptions. In effect, ID software makes it possible to easily integrate many components and submodels, but it does not check that the modeled causal relations are correct or meaningful. An ID and its output displays might look equally clear and convincing, whether or not it incorporates causal assumptions that correctly describe the real world. For example, in Figure 6.2, it would be easy to use a reduced-form linear regression model to populate the "slope of dose–response," assuming (perhaps incorrectly) that such a slope exists and is well defined, even if the true dose–response relation has very different slopes at different exposure concentrations (e.g., if it is J-shaped or U-shaped). To assess whether node-specific CPTs or regression models in an ID are causally interpretable requires careful analysis from outside the ID modeling process.

CONCLUSIONS: IMPROVING CAUSAL ANALYSIS OF HEALTH EFFECTS

The principles and techniques discussed in this chapter for modeling causal relations and for protecting against false conclusions and drawing sound causal inference from data are ready for practical use—and, indeed, are increasingly being used—in health effects research and risk assessment, both to reduce uncertainty about causality in concentration–response functions and to more clearly delineate needed distinctions between causal and noncausal associations. Taking seriously the need to apply more objective methods to assess causality in health risk assessment and public policy (and other) decision-making suggests the following policy-relevant perspectives:

- *Expert judgment-based assessments of causality and subjective causal interpretations of statistical associations are unreliable and prone to error and bias.* This is illustrated in examples where confidently expressed expert conclusions and more formal causal analyses conflict (as for several studies in Table 6.1). The prevalence of confirmation bias (Fugelsang et al., 2004; Gardner, 2009; Sunstein, 2009) makes it crucial for expert panels (or individuals) tasked with forming judgments about causation to seek out well-supported contrary views.

- *It is possible and practical to do better.* More objective methods for causal analysis are now readily available, and more informative designs and analyses (e.g., using panel data to study changes in exposure and response variables, instead of using regression models to study associations between their levels) can eliminate much of the speculation, controversy, and ambiguity surrounding causation in health effects research.

- *The credibility of conclusions about causation and the credibility of risk assessments and health benefits projections based on them should be assessed based on how well they provide sound, independently reproducible answers to specific, factual, causal questions.* These include addressing whether observed changes in hypothesized causal predecessors do in fact precede and help to explain or predict observed changes in their hypothesized effects. Passionate or confident beliefs about causation expressed by subject matter experts who have not yet addressed these questions using data and independently reproducible analyses should be regarded as expressions of personal belief rather than as answers to scientific questions.

Following these recommendations could transform health effects accountability research (Pope, 2010) by promoting health benefit estimates for exposure reductions that are more realistic, and more solidly based on reproducible science and data, than those driving headlines today. This would reduce needless controversies over the interpretation of ambiguous statistical associations, focus attention on the sizes of demonstrable real-world causal impacts, and shift the emphasis of health effects claims for emissions reductions toward more objective and independently verifiable risk analysis.

ACKNOWLEDGMENTS

This work was supported in part by a grant from the George Washington University Regulatory Studies Center in the Trachtenberg School of Public Policy and Public Administration. I thank Susan Dudley, Michael Greenberg, and Suresh Moolgavkar for excellent comments and suggestions that improved the exposition and that challenged me to think constructively about how to improve causal analysis. I also thank Goran Krstić for a close reading and useful discussions of air pollution epidemiology and causal interpretations of data. Portions of this chapter have appeared in L.A. Cox, Jr., *Improving Risk Analysis* (Springer, 2013) and in L.A. Cox, Jr., Improving causal inferences in risk analysis. *Risk Analysis* 2013 Oct;33(10):1762–71. doi: 10.1111/risa.12072.

REFERENCES

Albert I, Grenier E, Denis JB, Rousseau J. Quantitative risk assessment from farm to fork and beyond: A global Bayesian approach concerning food-borne diseases. Risk Anal. 2008;28(2):557–71.

Angrist JD, Pischke J-S. *Mostly Harmless Econometrics: An Empiricist's Companion.* Princeton University Press, Princeton, NJ, 2009.

Baio G, Pammolli F, Baldo V, Trivello R. Object-oriented influence diagram for cost-effectiveness analysis of influenza vaccination in the Italian elderly population. Expert Rev Pharmacoecon Outcomes Res. 2006;6(3):293–301.

Buka I, Koranteng S, Osornio-Vargas AR. The effects of air pollution on the health of children. Paediatr Child Health. 2006;11(8):513–6.

Burnside ES, Davis J, Chhatwal J, Alagoz O, Lindstrom MJ, Geller BM, Littenberg B, Shaffer KA, Kahn CE Jr. *Probabilistic computer model developed from clinical data in national mammography database format to classify mammographic findings.* Radiology 2009 Jun;251(3):663–72. Page CD, http://www.ncbi.nlm.nih.gov/pubmed/19366902 (accessed December 18, 2014).

Campbell DT, Stanley JC. *Experimental and Quasi-Experimental Designs for Research.* Rand McNally, Chicago, IL, 1966.

Carriger JF, Barron MG. Minimizing risks from spilled oil to ecosystem services using influence diagrams: The Deepwater Horizon spill response. Environ Sci Technol. 2011;45(18):7631–9.

Carriger JF, Newman MC. Influence diagrams as decision-making tools for pesticide risk management. Integr Environ Assess Manag. 2012;8(2):339–50.

Cheng J, Druzdzel MJ. AIS-BN: An adaptive importance sampling algorithm for evidential reasoning in large Bayesian networks. J Artif Intell Res. 2000;13: 155–188.

Chin K-S, Tang D-W, Yang J-B, Wong SY, Wang H. Assessing new product development project risk by Bayesian network with a systematic probability generation methodology. Expert Syst Appl Int J. 2009;36(6):9879–90.

Chuang KJ, Chan CC, Su TC, Lee CT, Tang CS. The effect of urban air pollution on inflammation, oxidative stress, coagulation, and autonomic dysfunction in young adults. Am J Respir Crit Care Med. 2007;176(4):370–6.

Clancy L, Goodman P, Sinclair H, Dockery DW. Effect of air-pollution control on death rates in Dublin, Ireland: An intervention study. Lancet. 2002;360(9341): 1210–4.

Cockburn G, Tesfamariam S. Earthquake disaster risk index for Canadian cities using Bayesian belief networks. Georisk. 2012;6(2):128–140.

Cox LA Jr. Regulatory false positives: True, false, or uncertain? Risk Anal. 2007;27(5): 1083–6.

Cox LA Jr. Hormesis for fine particulate matter (PM2.5). Dose-Response. 2012;10(2): 209–18.

Cox LA Jr, Popken DA, Ricci PF. Warmer is healthier: Effects on mortality rates of changes in average fine particulate matter (PM2.5) concentrations and temperatures in 100 U.S. cities. Regul Toxicol Pharmacol. 2013 Aug;66(3):336–46.

Daly R, Shen Q, Aitken S. Learning Bayesian networks: Approaches and issues. Knowl Eng Rev. 2011;26(2):99–157.

Das S, Lawless D, Ng B, Pfeffer A. Factored particle filtering for data fusion and situation assessment in urban environments. In Proceedings, 8th International Conference on Information Fusion Philadelphia, PA, July 25–28, 2005, (FUSION), Vol. 2, 955–962, IEEE Press, Piscataway, 2005.

Dash D, Druzdzel MJ. A note on the correctness of the causal ordering algorithm. Artif Intell. 2008;172:1800–08. http://www.pitt.edu/~druzdzel/psfiles/aij08.pdf (accessed August 25, 2014).

David LA, Wiggins CH. Benchmarking of dynamic Bayesian networks inferred from stochastic time-series data. Ann N Y Acad Sci. 2007;1115:90–101.

Druzdzel MJ, Simon HA. Causality in Bayesian belief networks. In UAI'93: Proceedings of the Ninth Annual Conference on Uncertainty in Artificial Intelligence, July 3–11, 1993, The Catholic University of America, Providence, Morgan Kaufmann Publishers, Inc., Washington, DC, 1993. http://www.sigmod.org/publications/dblp/db/conf/uai/uai1993.html (accessed December 18, 2014).

Eichler M, Didelez V. On Granger causality and the effect of interventions in time series. Lifetime Data Anal. 2010;16(1):3–32.

EPA (U.S. Environmental Protection Agency). The Benefits and Costs of the Clean Air Act from 1990 to 2020. Final Report—Rev. A. Office of Air and Radiation, Washington, DC, 2011.

Ezell BC, Bennett SP, von Winterfeldt D, Sokolowski J, Collins AJ. Probabilistic risk analysis and terrorism risk. Risk Anal. 2010;30(4):575–89.

Fann N, Lamson AD, Anenberg SC, Wesson K, Risley D, Hubbell. Estimating the national public health burden associated with exposure to ambient PM2.5 and ozone. Risk Anal. 2012;32(1):81–95.

Fenton NE, Neil M, Marquez D. Using Bayesian networks to predict software defects and reliability. J Risk Reliab. 2008;222(4):701–12.

Franklin M, Zeka A, Schwartz J. Association between PM2.5 and all-cause and specific-cause mortality in 27 US communities. J Expo Sci Environ Epidemiol. 2007; 17(3):279–87.

Freedman DA. Graphical models for causation, and the identification problem. Eval Rev. 2004;28(4):267–93.

Friede T, Henderson R, Kao CF. A note on testing for intervention effects on binary responses. Method Inform Med. 2006;45(4):435–40.

Friedman N, Goldszmidt M. Learning Bayesian networks with local structure. In Jordan MI, (Ed.), *Learning in Graphical Models*, 421–459. MIT Press, Cambridge, MA, 1998.

Friedman MS, Powell KE, Hutwagner L, Graham LM, Teague WG. Impact of changes in transportation and commuting behaviors during the 1996 Summer Olympic Games in Atlanta on air quality and childhood asthma. JAMA. 2001;285(7):897–905.

Fugelsang JA, Stein CB, Green AE, Dunbar KN. Theory and data interactions of the scientific mind: Evidence from the molecular and the cognitive laboratory. Can J Exp Psychol. 2004;58(2):86–95.

Gal Y, Pfeffer A. Networks of influence diagrams: A formalism for representing agents' beliefs and decision-making processes. J Artif Intell Res. 2008;33:109–147.

Gardner, D. *The Science of Fear: How the Culture of Fear Manipulates Your Brain.* Penguin Group, New York, 2009.

Gilmour S, Degenhardt L, Hall W, Day C. Using intervention time series analyses to assess the effects of imperfectly identifiable natural events: A general method and example. BMC Med Res Methodol. 2006;6:16.

Greenland S, Brumback B. An overview of relations among causal modelling methods. Int J Epidemiol. 2002;31(5):1030–7.

Gret-Regamey A, Straub D. Spatially explicit avalanche risk assessment linking Bayesian networks to a GIS. Nat Hazards Earth Syst Sci. 2006;6:911–26.

Guo H, Hsu WH. A Survey of Algorithms for Real-Time Bayesian Network Inference. AAAI/KDD/UAI-2002 Joint Workshop on Real-Time Decision Support and Diagnosis Systems. Edmonton, July 29, 2002.

Hack CE, Haber LT, Maier A, Shulte P, Fowler B, Lotz WG, Savage RE Jr. A Bayesian network model for biomarker-based dose response. Risk Anal. 2010;30(7):1037–51.

Harvard School of Public Health, Press release: "Ban on coal burning in Dublin cleans the air and reduces death rates" 2002. www.hsph.harvard.edu/news/press-releases/archives/2002-releases/press10172002.html (**accessed** August 25, 2014).

Health Effects Institute (HEI) Impact of Improved Air Quality During the 1996 Summer Olympic Games in Atlanta on Multiple Cardiovascular and Respiratory Outcomes. HEI Research Report #148. Jennifer L. Peel, Mitchell Klein, W. Dana Flanders, James A. Mulholland, and Paige E. Tolbert. Health Effects Institute. Boston, MA, 2010. http://pubs.healtheffects.org/getfile.php?u=564 (accessed August 25, 2014).

Hedley AJ, Wong CM, Thach TQ, Ma S, Lam TH, Anderson HR. Cardiorespiratory and all-cause mortality after restrictions on sulphur content of fuel in Hong Kong: An intervention study. Lancet. 2002;360(9346):1646–52.

Helfenstein U. The use of transfer function models, intervention analysis and related time series methods in epidemiology. Int J Epidemiol. 1991;20(3):808–15.

Howard RA, Matheson JE. Influence diagrams. Decis Anal. 2005;2(3):127–143.

Imberger G, Vejlby AD, Hansen SB, Møller AM, Wetterslev J. Statistical multiplicity in systematic reviews of anaesthesia interventions: A quantification and comparison between Cochrane and non-Cochrane reviews. PLoS ONE. 2011;6(12):e28422. doi:10.1371/journal.pone.0028422.

Ioannidis JPA. Why most published research findings are false. PLoS Med. 2005; 2(8):e124.

Jayasurya K, Fung G, Yu S, Dehing-Oberije C, De Ruysscher D, Hope A, De Neve W, Lievens Y, Lambin P, Dekker AL. Comparison of Bayesian network and support vector machine models for two-year survival prediction in lung cancer patients treated with radiotherapy. Med Phys. 2010;37(4):1401–7.

Kaufman JD. Air pollution and mortality: Are we closer to understanding the how? Am J Respir Crit Care Med. 2007;176(4):325–6.

Kelly F, Armstrong B, Atkinson R, Anderson HR, Barratt B, Beevers S, Cook D, Green D, Derwent D, Mudway I, Wilkinson P; HEI Health Review Committee. The London low emission zone baseline study. Res Rep Health Eff Inst. 2011;163:3–79.

Koller D, Milch B. Multi-agent influence diagrams for representing and solving games. Games Econ Behav, 2003;45(1):181–221.

Krstić G. Apparent temperature and air pollution vs. elderly population mortality in Metro Vancouver. PLoS One. 2011;6(9):e25101.

Lamm SH, Hall TA, Engel E, White LD, Ructer FH. PM 10 Particulates: Are they the major determinant in pediatric respiratory admissions in Utah County, Utah (1985–1989)? Ann Occup Hyg. 1994;38:969–972.

Lehrer J. Trials and errors: Why science is failing us. *Wired*. 2012. http://www.wired.co.uk/magazine/archive/2012/02/features/trials-and-errors?page=all (accessed January 28, 2012).

Lepeule J, Laden F, Dockery D, Schwartz J. Chronic exposure to fine particles and mortality: An extended follow-up of the Harvard six cities study from 1974 to 2009. Environ Health Perspect. 2012;120(7):965–70.

Li L, Wang J, Leung H, Jiang C. Assessment of catastrophic risk using Bayesian network constructed from domain knowledge and spatial data. Risk Anal. 2010;30(7): 1157–75.

Li L, Wang J, Leung H, Zhao S. A Bayesian method to mine spatial data sets to evaluate the vulnerability of human beings to catastrophic risk. Risk Anal. 2012;32(6): 1072–92.

Lipsitch M, Tchetgen Tchetgen E, Cohen T. Negative controls: A tool for detecting confounding and bias in observational studies. Epidemiology. 2010;21(3):383–8.

Liu RF, Soetjipto R. Analysis of three Bayesian network inference algorithms: Variable elimination, likelihood weighting, and Gibbs sampling. J Artif Intell Res. 2004. http://citeseerx.ist.psu.edu/viewdoc/download?doi=10.1.1.134.5771&rep=rep1&type=pdf (accessed December 18, 2014).

Liu F, Tian F, Zhu Q. Bayesian network structure ensemble learning. In *Advanced Data Mining and Applications: Lecture Notes in Computer Science*, 4632/2007:454–465. 2007. http://www.springerlink.com/content/043843r87730nn67/ (accessed August 25, 2014).

Lockamy III, A. Benchmarking supplier risks using Bayesian networks. Benchmarking Int J. 2011;18(3):409–27.

Lu T-C, Druzdzel MJ, Leong T-Y. Causal mechanism-based model construction. In UAI-2000: The Sixteenth Conference on Uncertainty in Artificial Intelligence, Stanford University, Stanford, CA, June 30–July 3, 2000, 353–62, Morgan Kaufmann Publishers, Inc., San Francisco, CA, 2000. http://www.auai.org/uai2000/www.cs. toronto.edu/uai2000/index.html (accessed December 18, 2014).

Maclure M. Multivariate refutation of aetiological hypotheses in non-experimental epidemiology. Int J Epidemiol. 1990;19(4):782–7.

Mansfield C, Sinha P, Henrion M. Influence Analysis in Support of Characterizing Uncertainty in Human Health Benefits Analysis. November, 2009. Final Report. Prepared for Amy Lamson U.S. Environmental Protection Agency Office of Air Quality Planning and Standards (OAQPS) Air Benefit and Cost Group (ABCG). EPA Contract Number EP-D-06-003. Research Triangle Park, NC, 27711. http://www. epa.gov/ttnecas1/regdata/Benefits/influence_analysis_final_report_psg.pdf (accessed December 18, 2014).

Marsh W, Bearfield G. Using Bayesian Networks to Model Accident Causation in the UK Railway Industry. In Probabilistic Safety Assessment and Management (PSAM7-ESREL'04): Proceedings of the 7th International Conference on Probabilistic Safety Assessment and Management, June 14–18, 2004, Berlin, 2004.

Meek C, Thiesson B. Probabilistic Inference for CART Network. Microsoft Research Technical Report MSR-TR-2010-40. Redmond, WA, 2010. http://research.microsoft. com/pubs/130854/CARTnetwork.pdf (accessed August 26 2014).

Mercer JB. Cold--an underrated risk factor for health. Environ Res. 2003;92(1):8–13.

Moore KL, Neugebauer R, van der Laan MJ, Tager IB. Causal inference in epidemiological studies with strong confounding. Stat Med. 2012;31(13):1380–404.

Morgan MG, Henrion M. *Uncertainty: A Guide to Dealing with Uncertainty in Quantitative Risk and Policy Analysis.* Cambridge University Press, New York, 1990, reprinted in 1998.

Neapolitan RE. The principle of interval constraints: A generalization of the symmetric Dirichlet distribution. Math Biosci. 1991;103(1):33–44.

NHS, Air pollution "kills 13,000 a year" says study. 2012. www.nhs.uk/news/2012/04april/Pages/air-pollution-exhaust-death-estimates.aspx (accessed August 2014).

Ottenbacher KJ. Quantitative evaluation of multiplicity in epidemiology and public health research. Am J Epidemiol. 1998;147:615–9.

Owens DK, Nease RF Jr. Development of outcome-based practice guidelines: A method for structuring problems and synthesizing evidence. Jt Comm J Qual Improv. 1993; 19(7):248–63.

Owens DK, Shachter RD, Nease RF Jr. Representation and analysis of medical decision problems with influence diagrams. Med Decis Making. 1997;17(3):241–62.

Pavlenko T, Chernyak O. Credit risk modeling using Bayesian networks. Int J Intell Syst. 2010;25(5):326–44.

Pearl J. An introduction to causal inference. Int J Biostat. 2010;6(2):Article 7.

Pelucchi C, Negri E, Gallus S, Boffetta P, Tramacere I, La Vecchia C. Long-term particulate matter exposure and mortality: a review of European epidemiological studies. BMC Public Health 2009 Dec 8;9:453.

Pope CA III. Respiratory disease associated with community air pollution and a steel mill, Utah Valley. Am J Public Health. 1989;79(5):623–8.

Pope CA. Accountability studies of air pollution and human health: Where are we now, and where does the research need to go next? 2010. http://pubs.healtheffects.org/getfile.php?u=584 (accessed August 26, 2014).

Powell H, Lee D, Bowman A. Estimating constrained concentration–response functions between air pollution and health. Environmetrics. 2012;23(3):228–37.

Regnier-Coudert O, McCall J, Lothian R, Lam T, McClinton S, N'dow J. Machine learning for improved pathological staging of prostate cancer: A performance comparison on a range of classifiers. Artif Intell Med. 2012;55(1):25–35.

Ren J, Jenkinson I, Wang J, Xu DL, Yang JB. A methodology to model causal relationships on offshore safety assessment focusing on human and organizational factors. J Safety Res. 2008;39(1):87–100.

Robins JM, Hernán MA, Brumback B. Marginal structural models and causal inference in epidemiology. Epidemiology. 2000;11(5):550–60.

Sarewitz D. Beware the creeping cracks of bias. Nature. 2012;485:149.

Smid JH, Heres L, Havelaar AH, Pielaat A. A biotracing model of Salmonella in the pork production chain. J Food Prot. 2012;75(2):270–80.

Stebbings JH Jr. Panel studies of acute health effects of air pollution. II. A methodologic study of linear regression analysis of asthma panel data. Environ Res. 1978;17(1): 10–32.

Sunstein, CR. *Going to Extremes: How Like Minds Unite and Divide.* Oxford University Press, New York, 2009.

Tebbens DRJ, Pallansch MA, Kew OM, Sutter RW, Aylward B, Watkins M, Gary H, Alexander J, Jafari H, Cochi SL, Thompson KM. Uncertainty and sensitivity analyses of a decision analytic model for posteradication polio risk management. Risk Anal. 2008;28(4):855–76.

Weidl G, Iglesias-Rozas JR, Roehrl N. Causal probabilistic modeling for malignancy grading in pathology with explanations of dependency to the related histological features. Histol Histopathol. 2007;22(9):947–62.

Wittmaack K. The big ban on bituminous coal sales revisited: Serious epidemics and pronounced trends feign excess mortality previously attributed to heavy black-smoke exposure. Inhal Toxicol. 2007;19(4):343–50.

Yim SH, Barrett SR. Public health impacts of combustion emissions in the United Kingdom. Environ Sci Technol. 2012;46(8):4291–6.

Yong E. Replication studies: Bad copy. Nature. 2012;485:298–300

Zhang NL. Probabilistic inference in influence diagrams, Computational Intelligence 1998;14(4):475–97.

CHAPTER 7

Making Decisions without Trustworthy Risk Models

Louis Anthony (Tony) Cox, Jr.

Cox Associates, NextHealth Technologies,
University of Colorado-Denver, Denver, CO, USA

CHALLENGE: HOW TO MAKE GOOD DECISIONS WITHOUT AGREED-TO, TRUSTWORTHY RISK MODELS?

How can risk analysts help to improve policy and decision-making when the approximately correct relation between alternative acts and their probable consequences is unknown? This practical challenge of *risk management with model uncertainty*, which was discussed in Chapter 4 from the standpoint of learning how to act while recognizing that future information may cause current models to be revised or replaced, arises in even more extreme form when learning by trial and error is prohibitively costly or potentially fatal. Yet, decision-making without the support of models that can be trusted to deliver at least approximately correct results is a frequent feature of modern life, in problems ranging from preparing for climate change to managing emerging diseases to operating complex and hazardous facilities safely.

This chapter reviews constructive methods for improving predictions and decisions when the correct description of the causal relation

Breakthroughs in Decision Science and Risk Analysis, First Edition.
Edited by Louis Anthony Cox, Jr.
© 2015 John Wiley & Sons, Inc. Published 2015 by John Wiley & Sons, Inc.

between decisions and outcome probabilities is unknown or is highly uncertain. These methods are not yet as familiar to many risk analysts as older statistical and model-based methods, such as the paradigm of identifying a single "best-fitting" model and performing sensitivity analyses for its conclusions. They provide genuine breakthroughs for improving predictions and decisions when the correct model is highly uncertain. We demonstrate their potential by summarizing a variety of practical risk management applications.

As a point of departure, we recognize that some of the most troubling risk management challenges of our time are characterized by *deep uncertainties*. Well-validated, trustworthy, risk models giving the probabilities of future consequences for alternative present decisions are not available; the relevance of past data for predicting future outcomes is in doubt; experts disagree about the probable consequences of alternative policies or, worse, reach an unwarranted consensus that replaces acknowledgment of uncertainties and information gaps with groupthink; and policy makers (and probably various political constituencies) are divided about what actions to take to reduce risks and increase benefits. For such risks, there is little or no agreement even about what decision models to use, and risk analysts may feel morally obliged not to oversimplify the analysis by imposing one (Churchman, 1967; Rittel and Webber, 1973). Passions may run high and convictions of being right run deep, in the absence of enough objective information to support rational decision analysis and conflict resolution (Burton, 2008).

Examples of risk management with deep uncertainties include deciding where, when, and how to prepare for future effects of climate change (and, perhaps, of efforts to mitigate it); managing risks from epidemics and new or deliberately spread pathogens; protecting valuable but vulnerable species, habitats, and ecosystems from irreversible loss; testing and reducing new interdependencies in financial systems to reduce risks of catastrophic failure; designing and managing power grids and energy and traffic networks to increase their resilience and reduce their vulnerability to cascading failures; and trying to anticipate and defend against credible threats from terrorists, cybercriminals, bank fraud, and other adversarial risks. The final section will return to these motivating challenges, after we have reviewed technical concepts and methods that can help to meet them.

		Level 1	Level 2	Level 3	Level 4
					Deep uncertainty
	Context	A clear enough future 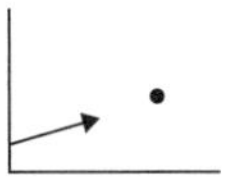	Alternate futures (with probabilities) 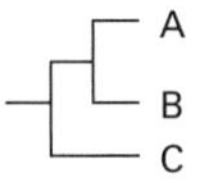	A multiplicity of plausible futures 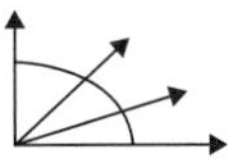	Unknown future 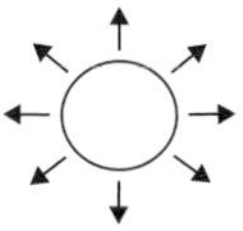
	System model	A single system model	A single system model with a probabilistic parameterization	Several system models, with different structures	Unknown system model; know we don't know
	System outcomes	A point estimate and confidence interval for each outcome	Several sets of point estimates and confidence intervals for the outcomes, with a probability attached to each set	A known range of outcomes	Unknown outcomes; know we don't know
	Weights on outcomes	A single estimate of the weights	Several sets of weights, with a probability attached to each set	A known range of weights	Unknown weights; know we don't know

(Left vertical axis label: **Determinism**; right vertical axis label: **Total ignorance**)

Figure 7.1   A suggested taxonomy of uncertainties (Walker, Marchau, and Swanson, 2010).

Figure 7.1 (Walker, Marchau, and Swanson, 2010) summarizes some uncertainties about matters of fact and value that separate deep uncertainties (right two columns, levels 3 and 4) from the more tractable uncertainties encountered in statistics and scenario analysis with known probabilities (left two columns, levels 1 and 2). (The "weights on outcomes" row at the bottom alludes to value weights and allows for uncertain preferences or utilities.)

Although these challenges are formidable, the underlying risks are too important to ignore and too complex to dispose of easily. Policy makers will continue to turn to risk analysts for help. Risk analysts, in turn, need to be familiar with the best available methods for improving risk management decisions under such trying conditions. This chapter summarizes recent progress in ideas and methods that can help. There has been great progress in technical methods for assessing and managing risks with deep uncertainties in recent years, usually using multiple models and scenarios. These methods are not yet widely used in risk analysis, compared to older methods that select a single statistical or simulation model and then perform sensitivity analyses on its results. The following sections seek to create an expository bridge from statistical methods and concepts that many risk analysts might already be familiar with (such as resampling techniques for robust statistical inference) to newer ideas from machine learning, robust optimization, and adaptive control that may be less familiar but that are promising for dealing with deep uncertainties in risk analysis.

PRINCIPLES AND CHALLENGES FOR COPING WITH DEEP UNCERTAINTY

There is no shortage of advice for managing risks with deep uncertainties. We should design fault-tolerant, survivable, and resilient organizations, systems, and infrastructures. We should experiment with possible improvements; learn quickly, effectively, and humbly from our own and others' mistakes and experiences (including accident precursors and unexpected events); and actively seek feedback and local "on the ground" information so that we can adapt flexibly to unforeseen circumstances and performance glitches. We should loosen or decouple the tight couplings and dependencies in existing complex systems and infrastructure—from oil rigs to financial systems—that set the stage for swiftly cascading failures and "normal accidents" (Harford, 2011). By adopting a vigilant, risk-aware mind-set and culture, we can, perhaps, build highly reliable organizations (HROs) around the five principles of preoccupation with failure, reluctance to simplify interpretations of data and anomalies, sensitivity to operations, commitment to resilience, and deference to expertise rather than to authority (Weick and Sutcliffe, 2007).

The practical problem is thus not finding logical principles for managing risks with deep uncertainties, but figuring out how best to implement them in detail. Risk analysts who, however rightly, respond to deep uncertainty by advocating greater learning and flexibility, or by promoting the virtues of adaptation and resilience to communities, institutions, and organizations, may be unsure how to bring them about, or how much good they will do if implemented. The following sections review methods that can help to improve risk management decisions when approximately correct models of the causal relation between controllable inputs and probabilities of valued (desirable or undesirable) outcomes are unknown, and learning and adaptation to new data are essential.

Point of Departure: Subjective Expected Utility Decision Theory

Traditional decision analysis and risk analysis make extensive use of models to predict the probable consequences of alternative risk management decisions. The paradigmatic analysis of decisions using subjective expected utility (SEU) theory, the gold standard for normative models of rational decision-making with level 1 uncertainties (see Chapter 2), proceeds as follows (Gilboa and Schmeidler, 1989):

- Identify a *choice set A* of alternative risk management acts. The decision problem is posed as choosing among the acts in A. Acts may represent not only alternative actions, such as resource allocations, but also rules for making decisions over time, such as alternative regulatory standards, adaptive feedback control policies, decision rules, collective choice rules, liability allocation rules, investment strategies, intervention trigger rules, and so on, depending on who is choosing what.

- Identify a set C of *possible consequences*. Choices of acts from A are to be made in an effort to make preferred consequences in C more likely and undesired ones less likely.

- *Quantify preferences.* This is typically done by assessing a von Neumann–Morgenstern utility $u(c)$, between 0 and 1, for each consequence c in C, such that the decision-maker is indifferent between receiving consequence c with certainty and receiving the most-preferred consequence in C with probability $u(c)$ and otherwise receiving the least-preferred consequence in C.

- *Optimize decisions.* Expected utility (EU) theory prescribes selecting an act in A that will maximize the expected value of $u(c)$, called EU. This prescription is typically justified by normative axioms for "rational" decision-making. It is implemented with the help of a probabilistic consequence (or risk) model, $\Pr(c \mid a)$, giving the probability of each consequence c if each act a is selected. Specifically, the EU of act a is $EU(a) = \sum_c \Pr(c \mid a)u(c)$, with the sum replaced by an integral if the consequence set is continuous.

- *Model and assess uncertainties.* If no well-validated empirical model, $\Pr(c \mid a)$, is available, then use subjective probability judgments to complete an SEU model. For example, suppose that the consequence of choosing act a depends on what else happens that is not directly controllable by the decision-maker (i.e., not in the choice set A). These other inputs—which, together with a, determine the consequence—lie in a set S of possible *scenarios* or *states* of nature. If $c(a, s)$ denotes the consequence that occurs when act a is chosen and state s occurs, then the EU of act a can be expressed as $EU(a) = \sum_s u[c(a, s)]\Pr(s)$. (More generally, if the consequence of a pair (a, s) is not deterministic, e.g., due to stochastic elements, then a conditional probability model for consequences, $\Pr(c \mid a, s)$, can be used to compute EU via the formula $EU(a) = \sum_c u(c)\Pr(c \mid a) = \sum_c u(c)[\sum_s \Pr(c \mid a, s) \Pr(s)]$.) If necessary, subjective probabilities $\Pr(s)$ for the states can be developed or elicited, for example, based on willingness to bet on each s compared to other events with known probabilities (perhaps after calibration training). SEU theory shows that a decision-maker with preferences satisfying certain axioms should behave as if she had coherent subjective probabilities $\Pr(s)$ and should choose acts that maximize EU calculated from these probabilities.

Thus, probabilistic consequence models, $\Pr(c \mid a)$ (perhaps built up from components, such as $c(a, s)$ or $\Pr(c \mid a, s)$ and $\Pr(s)$), play a crucial role in enabling rational decision-making via the SEU paradigm. A completely known EU *decision model* for supporting a risk management decision can be summarized by a quadruple $M = \{A, C, u(c), \Pr(c \mid a)\}$. When an approximately correct decision model is known and agreed on, EU provides a compelling normative framework for deciding what to do.

Four Major Obstacles to Applying SEU to Risk Management with Model Uncertainty

Fruitful though the SEU framework is, it cannot easily be applied to some of the most important risk management decisions that trouble modern societies, due to deep uncertainties. One obstacle to its practical application is uncertainty about *what alternatives are available* in the choice set A. A common problem is premature focusing on only a few salient options, which are not necessarily the best that could be devised. A second obstacle is uncertainty about the full range of *possible consequences* in *C*. The challenges of "unknown unknowns" or failures of imagination for potential consequences—for example, failure to correctly envision and account for all the important consequences that an act might make more probable or lack of confidence that all such important consequences have been identified—raise the concern that surprising "black swan" outcomes may occur, which were not considered when the decision was being made, but which would have changed the decision if they had been considered. The oft-bemoaned law of unintended consequences expresses this concern.

A third obstacle is that, even if *A* and *C* are both known, an *approximately correct risk model* Pr(*c* | *a*) for consequence probabilities for different acts may not be known (perhaps because the underlying state or scenario probabilities, Pr(*s*), are not well known). Different stakeholders may have conflicting beliefs about Pr(*c* | *a*) and hence conflicting beliefs about which act will maximize EU.

Finally, uncertainties or conflicts about *values and preferences* to be encoded in the utility function *u*(*c*) used to evaluate different consequences—for example, arising from differences in willingness to take risks to achieve potential rewards or because the preferences of future generations for consequences of current decisions are not well known—can make the expected utilities of different acts uncertain. Any of these obstacles can inhibit uncontroversial application of SEU theory to risk management problems, pushing a risk management problem to the right in Figure 7.1. If a completely known SEU model for supporting a risk management decision is denoted by $M = \{A, C, u(c), \text{Pr}(c \mid a)\}$, then the preceding difficulties can be viewed as instances of decision-making when the model, *M*, is unknown or disputed.

Decision-making without knowledge of, or agreement about, the basic assumptions needed to structure a decision problem by specifying a unique decision model, *M*, has been studied under headings such as deep uncertainty (Lempert and Collins, 2007), severe uncertainty (Ben Haim, 2001), model uncertainty, and wicked decision problems (Rittel and Webber, 1973). Constructive proposals to help guide risk management decision-making when relevant data are available, but a unique correct decision model is not known, are described next. Then we address the challenges of deeper uncertainty that arise when neither trustworthy predictive models nor relevant data are available at first, and it is necessary to learn and adapt as one goes. Finally, we will consider practical applications of these techniques.

TEN TOOLS OF ROBUST RISK ANALYSIS FOR COPING WITH DEEP UNCERTAINTY

Table 7.1 summarizes ten tools that can help us to better understand deep uncertainty and make decisions even when approximately correct models are unknown. They implement two main strategies: finding *robust decisions* that work acceptably well for many models (those in the uncertainty set) and *adaptive risk management* or learning what to do by well-designed and well-analyzed trial and error. Each is discussed in the following paragraphs, which also explain the different columns for generating, optimizing/adapting, and combining multiple model results.

Using Multiple Models and Relevant Data to Improve Decisions

When an approximately correct model linking acts to their probable consequences is unknown, but relevant data are available, good risk management decisions can often be made by combining predictions from multiple models that are consistent with available knowledge and data (e.g., as judged by statistical criteria discussed later). We will call the set of alternative models considered the *uncertainty set*.

A "good" decision, given the information available when it is made, can be defined as one to which no other choice is clearly preferable (e.g., by stochastic dominance (Buckley, 1986), giving clearly higher probabilities of preferred outcomes and lower probabilities of undesired outcomes, as assessed by all models in the uncertainty set). Alternatively, a "good" decision procedure might be defined as one that, despite all uncertainties, performs almost as well as some ideal procedure (e.g., optimal decision-making with perfect information or the best performing of all the models in the uncertainty set), as assessed in hindsight by the difference in rewards that they generate (often referred to as the regret for using the inferior model). Both approaches have led to strikingly successful procedures for using multiple models to let data inform decisions, even when no unique correct (or approximately correct) model is known. We will refer to both as methods for robust risk analysis, that is, risk analysis that delivers recommendations that are robust to deep (and other) uncertainties, especially about the correct probabilistic relation between acts and their consequences.

Table 7.1 Methods for decision-making with unknown models

Method	Model generation	Optimization/adaptation	Combination
Expected utility/SEU theory	One model specified	Maximize expected utility (over all acts in the choice set, A)	None
Multiple priors, models, or scenarios; robust control, robust decisions	Identify multiple priors (or models or scenarios), for example, all models close to a reference model (based on relative entropy)	Maximize the return from the worst-case model in the uncertainty set	Penalize alternative models based on their dissimilarity to a reference model
Robust optimization	Use decision-maker's risk attitude, represented by a coherent risk measure, to define the uncertainty set	Optimize objective function while satisfying constraints, for all members of uncertainty set	None
Average models	Use multiple predictive (e.g., forecasting) models	None	Simple average or weighted majority
Resampling	Create many random subsets of original data and fit a model to each	Fit models using standard (e.g., least squares, maximum likelihood) statistical criteria	Create empirical distribution of estimates
Adaptive boosting (AdaBoost)	Iteratively update training data set and fit new model	Reweight past models based on predictive accuracy	Use weights to combine models
Bayesian model averaging (BMA)	Include all models that are consistent with data based on likelihood	Condition model probabilities on data	Weight models by their estimated probabilities
Low-regret online decisions	Set of experts, models, scenarios, etc. is given, $\{M_1, M_2, \ldots, M_n\}$	Reduce weights of models that make mistakes	Weighted majority or selection probability
Reinforcement learning (RL) for MDPs: UCRL2	Uncertainty set consists of confidence region around empirical values	Approximately solve Bellman equations for most optimistic model in uncertainty set to determine next policy	Update from episode to episode based on new data
Model-free RL for MDPs: SARSA	No model used (model-free learning)	Approximately solve Bellman equations for unknown model	Update value estimates and policies based on new data

Several practical options are available for generating plausible models or scenarios (using various definitions of "plausible" or "consistent with data," as discussed in the following sections), optimizing decisions within and across these multiple possibilities, and combining the different decision recommendations into a final decision recommendation in a way that allows some performance guarantees for the quality of the result. The essence of robust risk analysis, for a large class of decision procedures, can be summarized as follows:

1. *Generate*: Generate or select multiple plausible models or scenarios, given available data and knowledge.
2. *Optimize/improve*: Find the best decision for each considered model or scenario. This may be interpreted as the decision "recommended" or "voted for" by that model or scenario. Alternatively, if optimization of decisions is not clearly defined or is not practical, but criteria and methods for improving models and decisions are available, then improve upon the ones considered so far until no further clear improvements can be made.
3. *Combine*: Use the multiple decision recommendations to recommend a final risk management decision by using some *combination rule* (such as majority voting) to combine the individual decision recommendations from step 2.

The robustness of the final decision recommendation can be defined and characterized in various ways, not only by the fraction of models that support it (or by upper bounds for the probability of models that do not) but also by upper bounds for the *difference in average reward* (e.g., EU or disutility) from following it *versus* from making the best decisions possible if the best model were known. The latter criterion leads to *low-regret* and *reinforcement learning* decision strategies for managing uncertain risks. The following paragraphs review methods for model generation, improvement, and combination to support robust risk analysis.

Robust Decisions with Model Ensembles

A crucial contribution to decision-making with deep uncertainty (Lempert and Collins, 2007; Bryant and Lempert, 2010) is the generation and analysis of many (e.g., thousands of) scenarios or models. Uncertainty about the most realistic decision model can be treated as just one more source of uncertainty, with each scenario in the uncertainty set now specifying a decision model to be used, as well as the values of other quantities that lie outside the choice set but that, together with the choice of act, affect consequences. If scenario probabilities are known, then EU can be maximized with respect to these probabilities. Even if the probabilities of different scenarios are not known, a decision that performs well by some criterion (e.g., which is undominated or which yields close to some provable upper bound on EU, given the information available when it is made) for most scenarios is likely to also do so in reality, if reality is well described by at least some scenarios in the uncertainty set and if this set is much more likely than the set of scenarios not considered—something that might be easier to assess than the individual scenario probabilities.

If one or a few decisions are "best" (e.g., maximizing scenario-specific expected utilities) or "good" for all or most of the considered scenarios, then these decisions are, in this sense, *robust to uncertainty* about which scenario in the uncertainty set is correct (if any). By contrast, if such ensemble analysis reveals that different choices are best for substantial fractions of the plausible scenarios, then it will be clear that no robust decision exists that makes the choice of decision immune to uncertainty about the correct scenario and that more information is therefore needed before a decision recommendation can be made that is robust, in this sense, to remaining uncertainties.

Example: Robust Decisions with Model Uncertainty

Tables 7.2 and 7.3 present two very different views of a risk management decision problem. In this example, a *perceived threat* of concern to some stakeholders (e.g., crop blights from climate change, genetically modified organisms in food, nanoparticles in air, electromagnetic radiation from cell phones, etc.) is assumed to have been identified, but it is not yet known whether complete scientific knowledge would reveal that

Table 7.2 Decision problem for an optimistic scenario

	Perceived threat is real, $p \leq 0.1$	Perceived threat is not real, $1 - p \geq 0.9$	Expected disutility
Act now	Disutility = 20	Disutility = 10	≥ 10
Wait for more information	Disutility = 40	Disutility = 0	≤ 4

Table 7.3 Decision problem for a pessimistic scenario

	Perceived threat is real, $p=0.4$	Perceived threat is not real, $1-p=0.6$	Expected disutility
Act now	Disutility $= 90$	Disutility $= 10$	$42 = 36 + 6$
Wait for more information	Disutility $= 100$	Disutility $= 0$	40

the exposures or activities of concern actually cause the harms that people worry about (abbreviated as "perceived threat is real") or not ("perceived threat is not real"). (More generally, one might be uncertain about the size of the threat, but these two states suffice to illustrate the basic challenge.) The alternative risk management acts being considered are to intervene now, perhaps by limiting exposures as a precautionary measure, or to wait for more information before deciding whether to intervene.

The tables show the expected disutility (scaled from 0 to 100) for each act–state pair. For simplicity, we assume that everyone agrees that the best choice of act is the one that minimizes expected disutility (equivalent to maximizing EU). However, perhaps due to the affect heuristic, optimistic stakeholders who think that the threat is probably not real ($p \leq 0.1$) also tend to think that its disutility, should it occur after all, will be modest (even though there is no logical reason that probability and severity must be positively correlated). Conversely, those who perceive the probability of a threat as being relatively high ($p=0.4$) also tend to perceive the severity of the threat (its disutility if it occurs) as being relatively great. Tables 7.2 and 7.3 are intended to capture these perceptions. Each constitutes one scenario.

The pessimists in Table 7.3 are shown as having crisp probabilities for the states (probability that the threat is real $= 0.4$), but the optimists in Table 7.2 have only imprecisely specified probabilities ($0 \leq p \leq 0.1$). Simple EU calculations show that acting now is less desirable than waiting, for the scenario in Table 7.2, if the threat probability is $p < 1/3$ (since then $20p + 10(1-p) < 40p$); hence, they prefer to wait; similarly, the pessimists described by Table 7.3 prefer to wait if $p < 1/2$, which it is ($p=0.4$ in this scenario). Hence, both scenarios prescribe waiting. Even if many other scenarios lie between these two extremes (i.e., with scenario-specific probabilities and disutilities lying between those in Tables 7.2 and 7.3), and even if we are ignorant of the respective probabilities of these scenarios, or even of what all the scenarios are, waiting for more information is a robust optimal decision with respect to this uncertainty set. (However, if the pessimists who see the world as in Table 7.3 become slightly less pessimistic, by changing their assessment of the disutility of acting now if the perceived threat is real from 90 to 80, then neither decision would be robustly optimal.)

Example: Robustness, Multiple Models, Ambiguous Probabilities, and Multiple Priors

EU theory has been extended to allow for uncertain or "ambiguous" probabilities and models and to consider ambiguity aversion as well as risk aversion. Instead of evaluating EU with respect to a unique "best-guess" prior probability distribution (or measure), an uncertainty set of multiple priors, all of which are considered plausible, can be used to represent ignorance of the true probability distribution. Then, axioms for decision-making with uncertain probabilities imply that a decision-maker should choose the act that *maximizes the minimum EU* obtained by using any of these plausible probability distributions (or measures) (Gilboa and Schmeidler, 1989). More generally, Maccheroni, Marinacci, and Rustichini (2006) presented conditions under which a decision-maker should choose the act in *A* that *maximizes the minimum penalized EU*, where different probability distributions or measures in the uncertainty set carry different penalties based on their plausibility. [Symbolically, such "variational preferences" prescribe choosing an act from choice set *A* to maximize the minimized value (over all members p of the uncertainty set) of the weighted sum $E_p[u(c \mid a)] + \alpha(p)$, where $E_p[u(c \mid a)]$ is the usual EU of act *a* if probability measure p is used to compute expected values and $\alpha(p)$ is the penalty for using p. ($\alpha(p) = 0$ if p is known to be correct and is larger for less plausible probability distributions or measures.) Robust decision-making (RDM) in this sense—maximizing the minimum expected reward (or credibility-penalized EU) over an uncertainty set of alternative probabilities—connects to a tradition of *robust control* in control engineering (Hansen and Sargent, 2001, 2008), in which controls are sought that perform well for all models not too dissimilar to a known reference model that is considered plausible but not necessarily correct. The measure of dissimilarity is typically based on information-theoretic metrics such as relative entropy or Kullback–Leibler divergence between the reference model and the model being weighted (Laeven and Stadje, 2011). Robust control of stochastic systems with somewhat misspecified models (not too dissimilar from the reference model) is mathematically equivalent to a special case of decision-making with multiple priors (Hansen and Sargent, 2008).

Example: Robust Optimization and Uncertainty Sets Using Coherent Risk Measures

One of the most useful paradigms for decision-making is *constrained optimization*, in which the choice set *A* consists of all values of one or more decision variables satisfying a set of constraints and the decision-maker seeks a set of values for the decision variables to maximize or minimize some objective function (e.g., average production of net benefits or average cost of losses per unit time, respectively). For example, the decision variables might be the amounts invested in risky stocks or opportunities, the constraint might be that the amount invested must not exceed a total budget available to invest, and the objective function might be the expected value of the resulting portfolio. More generally, a *robust linear optimization* problem (Bertsimas and Brown, 2009) seeks to maximize a weighted sum of decision variables (the linear objective function, e.g., the value of a risky portfolio) while keeping other weighted sums of the decision variables (e.g., the costs or resources required to implement the decision) acceptably small (the constraints), when it is only known that the values of the weights and constraints belong to some uncertainty set of alternative possibilities but the probabilities of different sets of weights and constraints are not known. Standard methods for solving deterministic constrained optimization problems, such as linear programming, which are suitable when the optimization problem is known with certainty, can give highly infeasible solutions when the problem data are uncertain; therefore, *robust optimization* methods must be used instead to address these model uncertainties (Ben-Tal, Ghaoui, and Nemirovski, 2009).

Any coherent risk measure representing the decision-maker's aversion to risk of violating a budget (or other linear) constraint can be expressed as an equivalent robust linear optimization problem with a convex uncertainty set that is derived directly from the coherent risk measure (Bertsimas and Brown, 2009). For example, if the conditional value-at-risk (CVaR) risk measure is used to specify that the expected value of cost in the worst (most costly) $x\%$ of cases must be no greater than some level b, then the corresponding uncertainty set can be generated by finding a set of probability measures that *represent* the CVaR measure of risk as minimizing expected values over that set. (Any coherent risk measure has such a minimum-expected-value-over-a-set-of-probabilities representation.) The uncertainty set for the corresponding robust optimization problem is then just a convex set (a polytope) of weighted averages of the probability measures that represent the coherent risk measure. The set of decisions that create "acceptable" risks of violating the linear constraint compared to the *status quo* according to a coherent risk measure is identical to the set of decisions that satisfy the constraint for all sets of weights in the uncertainty set.

Robust linear optimization problems can be solved via linear programming (due to the polytope shape of the uncertainty set). Both linear and nonlinear robust optimization problems can be computationally advantageous compared to nonrobust formulations, and the gap between the maximized EU or return from the correct model (if it were known) and the robust model is often surprisingly small (Ben-Tal, Bertsimas, and Brown, 2010; Bertsimas, Brown and Caramanis, 2011).

Averaging Forecasts

During the 1980s and 1990s, forecasting experts in time series econometrics and management science participated in several competitions (the "M-competitions") to discover empirically which forecasting models and methods worked best (e.g., minimizing mean squared error between forecast and subsequently revealed true values) in over 1000 different economic and business time series. One finding was that a simple arithmetic average of forecasts made by different methods usually out-performed any of the individual forecasts being averaged (Makridakis and Hibon, 2000). Averaging tends to reduce the error from relying on any single model (even the single best one), when even the best-fitting model is unlikely to be perfectly correct and even relatively poorly fitting models are likely to contribute some information useful for prediction. This is similar to Condorcet's centuries-old observation on majority voting with probabilistic knowledge: when each voter independently has a greater than 50% probability of correctly identifying which of two competing answers to a question is correct (assuming that one of them is), then majority rule in a large population of such voters has a probability close to 100% of selecting the correct answer—possibly very much greater than the probability for any of the individuals (Condorcet, 1785). Even if the voter opinions are not completely statistically independent, a similar conclusion often holds, as discussed later (e.g., for resampling, boosting, Bayesian model averaging (BMA), and online decisions). Note that this argument does not require knowing the probabilities that the different voters will be correct. Replacing voters with models and votes with model-based forecasts or probabilistic predictions provides heuristic motivation for the benefits of averaging predictions across multiple models.

Since these early experiments, a variety of *model ensemble methods* have been developed that seek to make predictions and decisions that are robust to some model uncertainties, in the sense that they work well for a large set of alternative plausible models and do not depend on assuming that any specific model (e.g., the best-fitting one) correctly describes or predicts the real situation.

Resampling Data Allows Robust Statistical Inferences despite Model Uncertainty

One way to generate multiple models to contribute to an ensemble prediction is to identify the "best" models (e.g., by traditional statistical criteria such as maximum likelihood or least squares or maximum a posteriori probability or minimum expected loss) for each of many randomly sampled subsets of the data. It is common in applied risk assessment that the correct statistical model for fitting a curve (e.g., a dose–response function) or estimating a quantity of interest (e.g., an odds ratio) from data is unknown. Then, modern computational statistical *resampling methods*—such as the bootstrap, jackknife, model cross-validation, and bagging—can create many random subsamples of the original data, fit a (possibly nonparametric) model or estimate to each subsample, and average these sample-specific estimates to obtain a final estimate (e.g., Molinaro, Simon, and Pfeiffer, 2005). The empirical distribution of the sample-specific estimates around the final estimate indicates how far from the final estimate the unknown true model might fall. Resampling can reduce bias from overfitting, leading to wider confidence intervals for model-based estimates (because model uncertainty is considered) and correspondingly fewer false positives for significant effects than selecting a single "best" model. It allows robust statistical inferences and model-based predictions, within limits (set in part by the model-fitting strategies used for the random samples, as well as by how the multiple samples are generated) even when the correct model is uncertain.

Adaptive Sampling and Modeling: Boosting

Instead of resampling data purely randomly, it turns out to be profoundly useful, for statistical classification problems, to construct deliberately biased samples that overweight data points that cannot yet be predicted well and then to iteratively improve models by fitting them to these deliberately biased training sets. On each iteration, a new statistical model is developed by fitting it to a new training set. Predictions from successive models are combined via a weighted-majority decision rule in which each model's "vote" (predicted class) is weighted based on its relative performance in correctly classifying data points in the training set. If the data points are then weighted based on how well they are predicted by the current best model, and these weights are used to determine the inclusion probability for each data point in the next training sample (with the least-well-predicted points receiving higher ("boosted") probabilities of being included), then a few hundred or thousand iterations can often generate an excellent statistical classifier, starting from even a weak initial predictive model that classifies data points with only slightly greater than random accuracy. Such *adaptive boosting* (AdaBoost) algorithms have proved highly successful in applications that require classifying cases into two or more classes. Examples include classification of credit applicants as "good" or "bad" credit risks (or into more than two credit risk categories) (Zhou and Lai, 2009), diagnosis of patients based on symptoms and markers (Tan, Chen, and Xia, 2009), prediction of which companies are most likely to go bankrupt over a stated time interval (Cortés, Gámez, and Rubio, 2007), prediction of toxicities of organic compounds (Su et al., 2011), and detection of intrusion in computer networks (Hu, Hu, and Maybank, 2008).

BMA for Statistical Estimation with Relevant Data but Model Uncertainty

One of the best-developed model ensemble methods is Bayesian Model Averaging (BMA) for statistical inference when the correct statistical model is uncertain. BMA seeks to weight model outputs (e.g., inferences, predictions, or decision recommendations) according to their probabilities of being correct, based on consistency with data. Like resampling methods, BMA creates many models (e.g., by considering all 2^n subsets of n candidate predictors in different regression models), but it weights each model based on its likelihood in light of the data, rather than fitting different models to different subsets of the data. (If there are too many plausible models to make it practical to generate and fit all of them, then sampling only those that are most consistent with the data, according to some statistical criterion, in the ensemble of considered models may yield a computationally tractable compromise.) BMA typically assesses consistency with the data by statistical criteria such as likelihood (i.e., model-predicted probability of the observed data) or likelihood penalized by model complexity, as reflected in degrees of freedom or number of constraints on data—the Bayesian information criterion (BIC). For example, a hypothesized causal model for multifactorial disease causation might be considered "consistent with data" if it implies a likelihood or BIC value for the observed data that is not much less than (e.g., is within an order of magnitude of) the maximum value for any model.

Given a model ensemble, BMA estimates the probability that a statistical property of interest holds (e.g., that a particular exposure variable is a significant predictor of a particular adverse health effect) or that a model-based prediction or conclusion is true (e.g., that the risk created by a given exposure exceeds a specified level), by considering the weighted fraction of the models in the ensemble that have that property or make that prediction, with each model weighted to reflect its conditional probability given the data (via a "Bayes factor" that reflects the likelihood of the data, given the model, in accord with Bayes' rule). An intuitive motivation is that the conditional probability that any conclusion, X, is true, given some set of observations that we will call Data, can be written (tautologically, via the law of total probability) as

$$\Pr(X|\text{Data}) = \Pr(X \mid M_1)\Pr(M_1|\text{Data})$$
$$+ \ldots + \Pr(X|M_n)\Pr(M_n|\text{Data})$$

where M_1, M_2, ..., M_n are any set of mutually exclusive and collectively exhaustive hypothesized models, *Data* represents any available observations, and $\Pr(M_j \mid \text{Data})$ is proportional to the likelihood of the data if model M_j is correct, $\Pr(\text{Data} \mid M_j)$. Various approximations made for computational tractability and convenience, such as only sampling from a large set of possible models and only considering models with tractable priors (glossed over in this brief overview) and with likelihood function values within an order of magnitude or so of the maximum-likelihood one, lead to different detailed BMA algorithms, appropriate for different types of statistical models ranging from regression models to Bayesian networks and causal graphs (Hoeting et al., 1999).

A substantial literature documents cases for which BMA-based statistical predictions or conclusions are less biased and more realistic than corresponding predictions or conclusions based on any single (e.g., best-fitting or maximum-likelihood) model. A typical result, as with resampling methods, is that confidence intervals for parameters estimated by BMA are wider, and type 1 errors (false positives) for falsely discovering what seem to be statistically "significant" results are correspondingly less common than when inferences are obtained from any single model, including the "best" model according to some model-selection criterion (Hoeting et al., 1999). This can have important implications for risk assessment results when model uncertainty is important. For example, when BMA is used to assess the statistical association between fine particulate matter (PM2.5) and mortality rates in some time series data sets, effects previously reported to be significant based on model selection (with model uncertainty ignored) no longer appear to be significant (Koop and Tole, 2004).

Learning How to Make Low-Regret Decisions

Resampling, boosting, and BMA methods are useful when they can fit multiple models to data that are known to be relevant for predicting future consequences of present decisions. If relevant data are initially unavailable, however, or if the relevance of past data to future situations is uncertain, then a different strategy is needed. This section considers what to do when data will be collected only as decisions are made and various different models (or experts or hypotheses or causal theories, etc.), with unknown probabilities of being correct, are available to inform decisions. This deeper uncertainty forces adaptive decision-making as relevant data become available, rather than predetermining the best course of action from available relevant data.

For systems with quick feedback, where the loss (or reward) for each act is learned soon after it is taken, some powerful approaches are now available for using multiple models to improve decisions. These situations can be modeled as *online decision* problems, in which what to do in each of many sequentially presented cases must be decided without necessarily knowing the statistical characteristics of the cases, which may be changing over time, or selected by one or more intelligent adversaries or influenced by continually adapting agents or speculators in a market.

Suppose that $\{M_1, M_2, ..., M_n\}$ are the different models (or theories, experts, scenarios, prediction algorithms, etc.) being considered, but their prior probabilities of being correct are unknown. Decision opportunities and feedback on resulting consequences arrive sequentially. For example, the risk manager may be confronted with a series of cases that require prompt decisions (such as stock market investment opportunities, patients to be treated, chemicals to be tested and classified, new drug applications or loan applications to be approved or rejected, etc.) If an approximately correct model were known, then it could be used to make decisions that would maximize the total reward earned from each decision, assuming that each choice of act for a case results in a consequence that can be evaluated by the decision-maker as having some value, which we call the "reward" for that decision in that case. In practice, an approximately correct model may not be known, but in online decision problems, the risk manager learns the actual consequence and reward soon after each decision; if the different models are specific enough, then the consequences and rewards that would have been received if each model had been used to make the decision may also be known.

The *cumulative regret* for using one model rather than another can be defined and quantified as the difference between the cumulative reward that would have been earned by following the decision recommendations from the second model instead of the first, if this difference is positive; equivalently, it is the cumulative loss from using the first model instead of the second. A good (or, more formally, *low-regret*) sequence of decisions, with respect to the ensemble $\{M_1, M_2, ..., M_n\}$, has an average regret per decision that approaches zero, compared to the best decisions that, in retrospect, could have been made using any of the models in the ensemble. In other words, a low-regret decision sequence does almost as well, on average, as if the decision-maker had always used the best model, as judged with the advantage of hindsight. Practical low-regret decision algorithms focus on homing in quickly on approximately optimal (or low-regret) decision rules, while keeping regret small during the learning period.

Somewhat remarkably, low-regret decision strategies are often easy to construct, even if the probabilities of the different models in the ensemble are unknown (Cesa-Bianchi and Lugosi, 2006). The basic idea is to weight each model based on how often it has yielded the correct decision in the past and to make decisions at any moment recommended by a *weighted majority* of the models. After each decision is made and its outcome is learned, models that made mistaken recommendations are penalized (their weights are reduced). Thus, the model ensemble produces recommendations that adapt to the observed performances of the individual models, as revealed in hindsight. (An alternative is to use the model weights to create *probabilities* of selecting each model as the one whose recommendation will be followed for the next case; such probabilistic selection (with the wonderful name of a "follow the perturbed leader" (FPL) strategy) also produces low-regret decision sequences (Hutter and Poland, 2005). A further variation (Blum and Mansour, 2007) is to adjust the weight on each model only when it is actually used to make a decision; this is important if the consequences that would have occurred had a different model been used instead are not known.) In each of these cases, weighted majority or FPL algorithms produce low-regret decision sequences; moreover, performance guarantees can be quantified, in the form of upper bounds for the average regret using the model ensemble algorithm compared to always using the best model (if it were known in advance).

If the environment is stationary (offering fixed but unknown probabilities of consequences for different decisions), then the low-regret strategies effectively learn, and then exploit, its statistical properties. If the environment changes over time, then low-regret strategies can be transformed to yield adaptive low *adaptive regret* strategies. These replace cumulative regret measures with measures of performance on successive intervals, to make the decision sensitive to changes in the underlying process (Hazen and Seshadhri, 2007). Risk analysts and policy analysts often recommend using *efficient adaptation* in light of future information to cope with deep uncertainty. Model ensemble decision algorithms provide one constructive framework to implement such recommendations.

Example: Learning Low-Regret Decision Rules with Unknown Model Probabilities

To understand intuitively how low-regret online decisions are possible, consider the extremely simple special case in which one must decide which of two possible decisions to make for each of a sequence of cases (e.g., invest or decline to invest in a new business opportunity, approve or deny a chemical product for consumer use, sell or hold a stock, administer or withhold an antibiotic in the treatment of a sick patient who might have a viral infection, etc.). After each decision is made, one of two possible outcomes is observed (e.g., business succeeds or fails, chemical product proves safe or hazardous, stock price moves up or down, patient would or would not have benefitted from the antibiotic, respectively). The decision-maker evaluates the results, assigning a "reward" (or loss) value to each outcome. The correct model for deciding what to do (or for predicting the outcome of each decision in each case) is uncertain. It belongs to some finite uncertainty set of alternative competing models $\{M_1, M_2, \ldots, M_n\}$ (perhaps developed by different experts or research groups or constituencies), but initially, the risk manager knows nothing more about which model is correct (e.g., there is no experience or available knowledge to assign meaningful probabilities to the individual models, or even to assign Dempster–Shafer beliefs to subsets of models, within the uncertainty set).

Despite this ignorance of the correct model, a low-regret sequence of decision can still be constructed as follows (Cesa-Bianchi and Lugosi, 2006):

1. Assign all the models in the uncertainty set the same initial weight, $1/n$.
2. As each case arrives, make the decision recommended by the weighted majority of models (i.e., sum the weights of all models in the ensemble that recommend each decision, and choose the decision with the maximum weight. In this simple example, with equal initial weights, this is the same as choosing the simple majority decision.) Resolve ties arbitrarily.
3. As long as the ensemble-based recommendation is correct (reward maximizing) for each case in hindsight, make no changes; but when the ensemble recommendation is mistaken, reduce the weights of all of the models that made the mistaken recommendation to zero.

Since majority rule is used, each new mistake eliminates at least half of the surviving models; thus, successive eliminations will lead to all decisions being made by the correct model (or to a subset of models that agree with the correct model), after a number of mistakes that is at most logarithmic in the number of models in the uncertainty set. After that, regret will be zero, and hence, average regret will approach zero as the correct model continues to be used.

For more realistic and complex cases, this simple procedure must be modified to achieve low-regret decisions. When there is no guarantee that the correct model is in the uncertainty set, and if only the consequences of the selected decisions are revealed (but not the consequences that other decisions would have produced), then the weights of models that contribute to incorrect (positive-regret) decisions are reduced only partially at each mistake, rather than jumping all the way to zero. Moreover, rather than making deterministic recommendations, the weights of models in the ensemble are used to set probabilities of selecting each possible act. Nonetheless, for a variety of sequential decision problems (including ones with more than two possible outcomes and more than two possible acts to choose among for each case), such refinements allow efficient adaptive learning of decision rules that perform almost as well on average as if the best model in the ensemble (as evaluated with 20–20 hindsight) were always used.

Reinforcement Learning (RL) of Low-Regret Risk Management Policies for Uncertain Dynamic Systems

The online risk management decision problems considered so far, such as deciding whether to approve loans, administer antibiotics, sell stocks, and so on, are perhaps less exciting than the grand challenges of risk management under deep uncertainty mentioned in the introduction. However, key ideas of low-regret decision-making can be generalized to a broad class of *RL* decision problems and algorithms that encompass many more complex risk management decision problems of practical interest. State-of-the-art RL algorithms also show how to generate continuous uncertainty sets based on observations and how to apply mathematical optimization to the resulting infinite ensembles of models to make low-regret decision in both stationary and changing environments.

Many risk management decision problems with deep uncertainties involve trading off relatively predictable immediate gains against uncertain future rewards or losses. Examples include extracting valuable nonrenewable resources with uncertain remaining reservoirs (such as oil or minerals); managing forests, vulnerable habitats, fisheries, or other renewable resources having uncertain population dynamics and extinction thresholds; attempted control of climate change with uncertain damage thresholds and points of no return; and medical use of antibiotics whose use increases, to an unknown extent, the risk of future antibiotic-resistant infections. In each case, a decision about how much benefit to extract now, given the present (perhaps uncertain) state of the word, yields an immediate reward, but it may also cause a transition to a new possibility inferior state offering different (perhaps lower or even zero) rewards for future actions.

For purposes of quantitative analysis, the usual formulation of such a problem is called a *Markov decision process* (MDP). In an MDP, choosing act a when the state of the system is s yields an immediate reward, $r(a, s)$, and also affects probabilities of transitions to each possible next state, $\Pr(s' \mid a, s)$, where $s' = $ a possible next state and $s = $ present state when act a is taken. (For stochastic rewards, the immediate reward may be the mean of a random variable with a distribution that depends on a and s.) A *decision rule* or *policy* for an MDP specifies the probability of taking each act when in each state. The set of such policies constitutes the choice set, A, in the standard EU formulation of decision theory discussed earlier, and the decision-maker seeks to identify the best policy. An *optimal policy* maximizes the *value* of the stream of rewards starting from each state; this value is usually denoted by $Q(s)$ and is defined as the expected sum of the immediate reward and the discounted value of future rewards, assuming that decisions now and in the future are consistently optimized. If β is the one-period discount factor, then optimal values, denoted by $Q(s)$, satisfy the following equation (the Bellman equation):

$$Q(s) = \max_{a \in A} \{r(a,s) + \beta Q(s') \Pr(s' \mid a,s)\}$$

In words, the optimized reward starting from state s is the maximized (over all possible current acts) sum of the immediate reward plus the maximized expected discounted future reward starting from the next state. This system of equations (one for each s) can be solved for the optimal policy, by standard algorithms from operations research (such as linear programming, value iteration, policy iteration, and stochastic dynamic programming) or by RL algorithms (such as Q-learning or temporal difference learning) that use successive empirical estimates of the optimal value function, based on the observed history of states, acts, and rewards so far, to gradually learn an optimal or nearly optimal policy (Sutton and Barto, 2005).

Robust low-regret risk management policies for MDPs (Regan and Boutilier, 2008) generate low regrets even when the reward distributions and state transition probabilities are initially not known, but must be estimated from observations, and even when they may change over time, rendering what has been learned so far no longer useful. These complexities move the decision toward the right in Figure 7.1—the domain of deeper uncertainties. Practical applications of RL algorithms to date have ranged from controlling hazardous chemical production processes to maximize average yield under randomly changing conditions while keeping the risk of entering dangerous process states within specified bounds (Geibel and Wysotzk, 2005) to devising stoplight control policies to reduce jams and delays in urban traffic (Gregoire et al., 2007).

Experiments and brain imaging (functional MRI) studies of human subjects suggest that RL also has neural correlates, with the human brain processing differences between anticipated and obtained rewards for different policies under risk, and subsequently adapting perceptions and behaviors, in ways that can be interpreted in terms of RL algorithms (e.g., Kahnt et al., 2009). For example, whether subjects successfully learn which of four risky reward processes generates the highest average, based on repeated trial-and-error learning, appears to be predicted by the strength of physiologically measurable signals involved in RL (Schönberg et al., 2007), although other experiments show that learning is also affected by mental models (possibly incorrect) of processes generating data (Green et al., 2010).

Example: RL of Robust Low-Regret Decision Rules

If a decision-maker must make choices in an unknown MDP model, with only the sets of possible states and acts (S and A) known, but rewards and state transition probabilities resulting from taking act a in state s having to be estimated from experience, then a low-regret strategy can be constructed using the following principle of *optimism in the face of uncertainty* (Jaksch, Ortner, and Auer, 2010):

1. Divide the history of model use into consecutive *episodes*. In each episode, a single policy is followed. The episode lasts until a state is visited for which the act prescribed by the current policy has been chosen as often within the current episode as in all previous episodes. (The new episode thus at most doubles the cumulative number of occurrences of any state–act pair.) When an episode ends, the data collected is used to update the uncertainty set of considered models, as well as the policy to be followed next, as described next.

2. At the start of each episode, create a new uncertainty set of plausible MDP models from confidence intervals around the empirically observed mean rewards and transition probabilities.

3. Choose an optimistic MDP model (one yielding a high average reward) from the uncertainty set. Solve it via operations research optimization techniques to find a near-optimal policy.

4. Apply this policy until the episode ends (see step 1). Then, return to step 2.

The analysis of a detailed algorithm (UCRL2, for upper confidence RL) implementing these steps shows a high probability (depending on the confidence levels used in step 2 to generate uncertainty sets) of low regret, compared to the rewards that would have been achieved if optimal policies for each of the true MDPs had been used (Jaksch, Ortner, and Auer, 2010). This result holds when any state can be reached from any other in finite time by appropriate choice of policies, and even when the true but unknown underlying MDP (i.e., reward distributions and transition probabilities) can change at random times (or in any other way that is oblivious to the decision-maker's actions), provided that the number of changes allowed in an interval is finite. Intuitively, the UCRL2 algorithm seeks the best return by exploring different plausible models, starting with those that would yield the best returns if correct. As data accumulates, confidence intervals around estimated model parameters shorten. When the current model no longer appears best, exploration switches to a different model. The UCRL2 algorithm learns efficiently and can adapt to changes in the underlying unknown MDP quickly enough so that the policies it recommends are unlikely to spend long yielding returns much lower than those from the best policies given perfect information.

Example: Model-Free Learning of Optimal Stimulus–Response Decision Rules

Rather than solving the Bellman equations directly, RL algorithms use data to approximate their solution increasingly well. For example, the *state–act–reward–state–act* (SARSA) RL algorithm updates the estimated value (the sum of immediate and delayed rewards) from taking act a in state s, denoted by $Q(s, a)$, via the equation

$$\text{New } Q(s,a) \text{ value} = \text{previous } Q(s,a) \text{ value}$$
$$+ \alpha[\text{change in estimated value of } Q(s,a)]$$

where α is a *learning rate* parameter and the change in the estimated value of $Q(s, a)$ is the difference between its new value (estimated as the sum of the most recently observed immediate reward and the previously estimated discounted value starting from the observed new state) and its previously estimated value:

$$[\text{Change in estimated value of } Q(s,a)]$$
$$= [r(s,a) + \beta Q(s',a')] - Q(s,a).$$

(Here, a' is the act taken in the observed next state, s', according to the previously estimated value function $Q(s, a)$; and $r(s, a) + \beta Q(s', a')$ is the estimated value just received when act a was taken in state s.) The difference between this estimate of value just received and the previous estimated value $Q(s, a)$ expected from taking act a in state s provides the feedback needed to iteratively improve value estimates and resulting policies. The change in the estimated value of $Q(s, a)$ is zero only when its previously estimated value agrees with its updated value based on the sum of observed immediate reward and estimated delayed reward starting from the observed next state, that is, only when $Q(s, a) = r(s, a) + \beta Q(s', a')$. When this condition holds for all states, the Bellman equation is satisfied, and the observed sequence of SARSA data (s, a, $r(s, a)$, s', a') has been used to learn the optimal policy. Detailed implementations of this idea (e.g., incorporating randomization to assure that all act–state pairs will eventually be tried with nonzero probability and specifying the act to be selected in each state, typically as the "epsilon-greedy" one that chooses an act at random with small probability and otherwise chooses the one that maximizes the current estimated expected value of $r(s, a) + \beta Q(s', a')$, perhaps with statistical regression or nonparametric smoothing models and Monte Carlo simulation of a random sample of future trajectories used to approximate $Q(s', a')$ for large state spaces) yield practical RL algorithms for a variety of sequential decision problems with random transitions and immediate and delayed losses or rewards (Szepesvari, 2010).

Many RL algorithms learn by comparing the estimated rewards received using the current policy to the best estimated rewards that could have been received (as predicted by a model) had a different policy been used instead and revising the current policy based on this difference (which can be interpreted as a measure of regret). By contrast, the SARSA algorithm uses only the *observed data* on what was done and what reward was experienced (the SARSA data) to update the value estimates for act-state pairs and to gradually learn an optimal policy. No model of the underlying MDP (or other process) is required. In effect, the learner maintains estimated values for an ensemble of different stimulus–response (i.e., act-state) pairs, updates these value estimates based on the experienced differences between obtained and expected rewards, and uses them to decide what to do as each new state occurs. Such adaptive learning is suitable even when no model is available and will converge to the optimal policy for the underlying MDP, if one exists, under quite general conditions, even if the unknown MDP itself occasionally changes (Yu, Mannor, and Shimkin, 2009).

Recent work has started to extend RL algorithms to *partially observable* MDPs (POMDPs) in which the state at each moment (e.g., the size of a fishery stock) is not known with certainty, but must be inferred from statistical information (e.g., sampling). State-of-the-art RL algorithms for POMDPs balance *exploration* of new or underinvestigated decision rules (each of which maps histories of observed information, acts, and rewards to decisions about what act to take next) and *exploitation* of known high-performing decision rules. Similar to SARSA, this approach can learn optimal or near-optimal polices for the underlying POMDP, if one exists, even without a model of the process (Cai, Liao and Cari, 2009; Ross et al., 2011).

Ongoing extensions and refinements of these ideas—especially *multiagent (social) learning* and *evolutionary optimization* algorithms, in which the (perhaps fatal) experiences of some agents help to inform the subsequent choices of others (Waltman and van Eck, 2009)—will bring further improvements in the ability to solve practical problems. However, the techniques summarized in Table 7.1 already suffice to support many valuable applications.

APPLYING THE TOOLS: ACCOMPLISHMENTS AND ONGOING CHALLENGES FOR MANAGING RISKS WITH DEEP UNCERTAINTY

Conceptual frameworks and technical tools such as those in Table 7.1 have practical value insofar as they help to improve risk management decisions with deep uncertainties. This section sketches applications of robustness and adaptive risk management methods to practical risk management problems with deep uncertainties and highlights some key challenges.

Before seeking sophisticated solutions to difficult problems, of course, it is well to cover the basics: pay attention to what doesn't work, and stop doing it; if possible, encourage many independent experiments on a small scale to find out what works better; identify, reward, and spread successes; and don't bet too heavily on unvalidated models or assumptions (Harford, 2011). The increasing capabilities of technical methods should not lead to neglect of such useful commonsense advice.

Planning for Climate Change and Reducing Energy Waste

In *robust decision making* (RDM), participants develop multiple scenarios—perhaps with the help of computer-aided scenario generation and an experienced facilitator (Bryant and Lempert, 2010)—to identify potential vulnerabilities of proposed decisions, such as where to build a road to connecting villages. These scenarios help participants to identify cost-effective ways to change the proposed decision to decrease vulnerabilities (e.g., potential loss of the road due to flooding or mud slides) and to develop increasingly robust decision options.

RDM has been advocated as a practical way to help multiple stakeholders in communities and developing countries engage in planning for climate change and infrastructure development (Lempert and Kalra, 2008). Some limitations are that a robust decision may not exist, and the most relevant and likely scenarios, as viewed in hindsight, may not be identified during planning. (For example, empirical surprises, such as larger-than-predicted effects of "global dimming," might not be considered among the scenarios, leading to an ensemble of predictions with uncertain or debated credibility (Srinivasan and Gadgil, 2002).) However, practical experience suggests that RDM can be helpful in envisioning and planning for possible futures (Bryant and Lempert, 2010).

While scenario-based planning methods such as RDM can help plan large-scale adaptation to envisioned potential changes, adaptive risk management methods can also guide smaller, immediate changes that significantly reduce energy waste and pollution by increasing the efficiency of energy consumption in uncertain environments. For example, RL algorithms have been used to design more efficient building energy conservation programs (subject to comfort constraints) (Dalamagkidis et al., 2007), devise more efficient use and coordination of stop lights to greatly reduce time spent by vehicles in urban traffic (Balaji, German, and Srinivasan, 2010), and optimize dynamic power used by devices (Wang, Xie, and Ammari, 2011). These applications reduce energy consumption without decreasing quality of life, by adaptively reducing wastes of energy.

Sustainably Managing Renewable Resources and Protecting Ecosystems

Sustainable management and harvesting of renewable resources can be formulated in terms of MDPs (or generalizations, such as semi-MDPs, in which the times between state transitions may have arbitrary distributions, or POMDPs). When the resources extend over large areas, with subareas developing differently over time, then the spatially distributed control problem of managing them can be factored into many local MDPs, represented as the nodes of a network, with local dependencies between the MDPs indicated by edges between nodes. Such graph-based MDPs (GMDPs) represent a variety of spatially distributed control problems in forestry and agriculture (Forsell, Garcia, and Sabbadin, 2009).

As an example, in a large commercial forest consisting of many stands of trees, a decision must be made about when to harvest each stand, taking into account that random severe windstorms (perhaps every few decades) pose a risk of wiping out most of the commercial value of a stand that is blown down before it is harvested but that neighboring stands can provide some shelter to each other and hence reduce risk of wind damage (Forsell, Garcia, and Sabbadin, 2009). If the probability distributions for rewards (e.g., based on market values of the crop over time) and state transition probabilities (e.g., based on statistics for windstorm arrival times and severities) were known in advance (level 1 uncertainty), then a state-of-the-art way to devise a value-maximizing harvesting policy would be to use *simulation–optimization*. As discussed in Chapter 3, simulation-optimization tries one or more initial policies (perhaps a mix of randomly generated and historical ones), simulates the consequences of each policy many times via Monte Carlo simulation using the known probability distributions, and iteratively improves policies until no further increases in the reward (e.g., average simulated net present value) can be found. Coupled with design-of-experiment principles for adaptively exploring the set of policies, together with sophisticated optimization steps (e.g., evolutionary optimization routines), current simulation–optimization algorithms can solve a wide range of forestry management problems under level 1 uncertainty. These include multicriteria decisions in which the utility derived from biodiversity, carbon sequestration, and standing forests, as well as the market value of timber, is taken into account (Yousefpour and Hanewinkel, 2009).

Simulation–optimization is impossible under deep uncertainty, however, because the probability distributions of consequences for different policies are unknown. Instead, current algorithms for risk management of GMDPs with unknown probabilities use collaborative multiagent RL algorithms. Each "agent" (typically identified with one node of the GMDP) makes decisions about one part of the problem (e.g., when to harvest one specific stand in a commercial forest). Each agent must coordinate with its neighbors to achieve optimal results. This is well within the capabilities of current multiagent RL algorithms for spatially distributed management of agricultural and forest resources (Forsell, Garcia, and Sabbadin, 2009).

Similar RL algorithms have been developed to adaptively manage risks of forest fires, which again pose locally linked risks that increase with time since last harvest (Chades and Bouteiller, 2005), and to protect and conserve biodiversity in Costa Rican forests over time, by adaptively coordinating and optimizing the reservation of subareas that will not be commercially exploited, in order to preserve habitats and species (Sabbadin et al., 2007). POMDPs are now starting to be used to optimize allocation of scarce conservation resources to multiple conservation areas, when the presence and persistence of threatened species in each area is uncertain (McDonald-Madden et al., 2011). Thus, current applications of RL can help to protect forests and other ecosystems, as well as to manage commercial forests and other resources over long periods in the presence of uncertain, and possibly changing, risks.

Managing Disease Risks

Like the spatial spread of wind damage, forest fires, and habitat loss or gain, many contagious diseases also have strong spatial, as well as temporal, dependencies. Stopping the spread of an epidemic requires deciding not only how to act (e.g., vaccine vs. quarantine) but also where and when and with what intensity. The stakes are high: failing to quickly contain a potential epidemic or pandemic can impose enormous economic and health costs. For example, one recent estimate of the economic consequences of delaying detection of a foot-and-mouth disease (FMD) outbreak in a California cattle herd from 7 days to 22 days is about $66 billion (with over half a billion of additional loss and 2000 additional cattle slaughtered for each extra *hour* of delay after 21 days) (Carpenter et al., 2011). Managing such risks in real time, with constantly changing spatiotemporal disease data and uncertainties about where and when new cases may be discovered, requires a new generation of risk management tools to inform intervention decisions far more quickly than traditional methods. RL algorithms are being developed to meet this need.

For several decades, simulation–optimization has been applied to design epidemic risk management plans for both animal and human contagious diseases, when infectious disease control models (e.g., for mass dispensing of stockpiled medical countermeasures) involve only level 1 or level 2 uncertainties (Lee et al., 2010). For epidemic models with deeper uncertainties, RL optimization of policies is now starting to be used. For example, RL algorithms applied to a stochastic simulation model of the spread of an H1N1 influenza pandemic and its consequences—from illnesses and deaths to healthcare expenses and lost wages and to shortages of vaccines, antiviral drugs, and hospital capacity—have recently been proposed to coordinate and optimize risk mitigation measures (early response, vaccination, prophylaxis, hospitalization, and quarantine applied at different times and locations) to create a cost-effective overall risk management strategy (Das, Savachkin, and Zhu, 2007).

In livestock, the spread of the highly contagious FMD can be controlled by a combination of vaccination and culling. Both overreaction and underreaction cost animal lives and economic losses; therefore, adroit and flexible risk management that exploits information as it becomes available is very valuable. Recent research suggests that adaptive risk management of FMD epidemics substantially outperforms traditional prespecified control strategies (in which observed cases trigger automatic culling and/or vaccination within a set area around affected farms), saving unnecessary loss of animal life and more quickly suppressing FMD (Ge et al., 2010).

Robust, ensemble, and adaptive risk management techniques are also starting to be used to improve medical screening, diagnosis, prediction, and treatment of a variety of diseases. Examples include the following:

- *Earlier detection of Alzheimer's.* Ensemble prediction methods can dramatically improve ability to detect and predict some medical conditions from data. The challenging task of using brain imaging data to automatically identify women with mild Alzheimer's disease is one where AdaBoost appears to substantially improve accuracy (Savio et al., 2009), and detection of Alzheimer's in brain MRIs by model ensemble methods that incorporate AdaBoost compares favorably even to manually created "gold standard" classifications (Morra et al., 2010).
- *Improving HIV treatment using RL.* A model-free RL algorithm has been proposed for using clinical data to decide adaptively when to cycle HIV patients off of harsh drug therapies, as part of a structured treatment interruption program designed to reduce risk of acquisition of drug resistance, as well as alleviating side effects (Ernst et al., 2006). The RL algorithm works directly with clinical data (e.g., observed levels of CD4+ T cell counts), with no need for an accurate model of HIV infection dynamics.
- *Treating depression.* RL algorithms that estimate value functions (the Q functions in the Bellman equation) despite missing data (e.g., caused by incomplete compliance and nonresponse bias in the patient population) have been used to adaptively refine treatments of

depressed patients by adjusting the combination of antidepressants administered over time, based on patient responses, to achieve quicker and more prevalent relief of symptoms (Lizotte et al., 2008).

- *Managing ischemic heart disease (IHD) and other dynamic diseases.* The problems of managing various dynamic diseases over time based on inconclusive observations have been formulated as MDPs and POMPDs (e.g., Schaefer et al., 2004; Alagoz et al., 2011). For example, for IHD, the physician and patient must decide when to administer or change medication, schedule stress tests or coronary angiograms, perform angioplasty or coronary artery bypass graft surgery, and so on, based on time-varying information of uncertain relevance that may range from reports of chest pain to EKG readings. This disease management process has been formulated as a POMD (Hauskrecht and Fraser, 2000), and uncertainty sets and practical solution algorithms for imprecisely known POMDs have been developed (Itoh and Nakamura, 2007; Ni and Liu, 2008).

- *Optimizing treatment of lung cancer patients in clinical trials.* Treatment of patients with advanced lung cancer typically requires switching among different lines of chemotherapy. RL algorithms are now being developed to approximately optimize the treatment of individual patients even when not enough is known to model the progression of cancers in detail (Zhao, Kosorok, and Zeng, 2009). The authors note that "reinforcement learning has tremendous potential in clinical research because it can select actions that improve outcomes by taking into account delayed effects even when the relationship between actions and outcomes is not fully known."

- *Predicting toxicity of chemicals.* Ensemble learning and prediction methods, including AdaBoost and its generalizations, have recently been shown to improve prediction of mechanisms of toxicity for organic compounds (e.g., phenols) based on molecular descriptors (Niua et al., 2009) and to out-perform other QSAR methods (Svetnik et al., 2005).

- *Better targeting of radiation therapy under uncertainty.* Robust optimization of intensity-modulated proton beam therapy spares more healthy tissues and organs than conventional optimization methods (e.g., based on probabilistic margins of error) while providing excellent coverage of the target tissue despite range and setup uncertainties (Fredriksson, Forsgren, and Hårdemark, 2011; Inaniwa et al., 2011).

Multiobjective evolutionary optimization algorithms have also been developed to automatically identify undominated choices for beam angles and intensities in radiation therapy treatment planning (Fiege et al., 2011).

• *Reducing schizophrenia hospitalization episodes.* Model ensemble predictors incorporating AdaBoost have been used recently to improve prediction of schizophrenia relapses in patients participating in a weekly remote patient monitoring and disease management program (via a PC-to-phone platform), increasing specificity of predictions from 0.73 to 0.84 while keeping sensitivity at 0.65 (Hrdlicka and Klema, 2011).

These examples suggest the potential for robust and adaptive methods to improve health risk management under uncertainty. This potential is only starting to be realized, since the methods are still relatively new, but it seems certain that many more practical applications in medical decision and risk analysis will be seen over the next few years.

Maintaining Reliable Network Infrastructure Service Despite Disruptions

Quickly containing and recovering from cascading failures in a power grid is somewhat analogous to quickly suppressing a spreading epidemic. In both, observations and control opportunities are spatially distributed; costly preemptive measures can be taken at different places (e.g., vaccinating as-yet uninfected flocks or shedding power loads before generators are knocked off grid); and a quick, effective response can potentially avert orders-of-magnitude larger losses. It is therefore perhaps unsurprising that multiagent reinforcement learning (MARL) algorithms (especially hierarchies and teams of RL controllers, each using an RL algorithm) are now being studied as effective risk management tools for increasing network resilience and responding to catastrophic failure events. For example, a two-level hierarchical control framework has recently been proposed to manage power generation and distribution in interconnected power grids under changing load and hydrothermal energy supply conditions (Zhou, Chan, and Yu, 2011). Model-free RL (via Q-learning) is used both to figure out how best to implement high-level commands at generation units and what high-level commands to give them to meet changing demands reliably and cheaply across the interconnected areas under normal conditions.

In the event of a catastrophic failure event that disables one or more generators (e.g., a storm, accident, or attack), decentralized (multiagent) Q-learning can again be used to quickly detect and prevent cascading failures and rapidly restore power grid systems (Ye, Zhang, and Sutato, 2011). Under such a contingency, adaptive load shedding, that is, selective deliberate dropping of electric power, keeps the network stable, preventing the spread of blackouts and minimizing power losses to customers as failures are isolated, power is rerouted, and service is automatically restored (Jung et al., 2002).

Similarly, multiagent distributed RL algorithms facilitate quick automated rerouting of data packet traffic in telecommunications networks following loss of fibers or switching centers, helping to make these networks highly resilient to equipment and link failures. Although vehicles cannot be rerouted as easily as data packets or electric power, control of urban traffic flow by applying similar distributed RL algorithms to traffic lights can reduce average delays and expedite passage of emergency equipment when traffic networks and communications networks are interlinked (Kuyer et al., 2008).

Adversarial Risks and Risks from Intelligent Agents

Methods of ensemble, robust, and adaptive risk analysis do more than provide useful concepts and detailed algorithms for coping with model uncertainty (including ambiguous beliefs and preferences) in a variety of practical applications. They also shed light on some key theoretical questions in risk analysis, for example, by providing performance guarantees for how quickly adaptive low-regret risk management policies learned from data converge to approximately the best possible policy or by giving upper bounds on the size of the cumulative difference in rewards obtained from the policy used versus those that would have been obtained from the perfect-information optimal policy or some other reference policy. Mathematical analysis shows that risks from intelligent adversaries cannot necessarily be managed effectively by using the same concepts and methods as for risks from nonintelligent sources: the same performance guarantees do not hold for systems that respond intelligently to a decision-maker's choices as for systems that do not (Yu, Mannor, and Shimkin, 2009).

This does not mean that the methods are not useful for detecting and mitigating vulnerabilities to deliberate attacks. Indeed, RL algorithms for POMDPs have been shown to improve the performance of early detection systems for anthrax outbreaks and proposed for use in reducing the consequences of possible bioterrorist attacks (Izadi and Buckeridge, 2007). RL algorithms are also used successfully to detect fraud in health insurance and auto insurance data (see background in Bolton and Hand, 2002; Lu, Boritz, and Covvey, 2006), and cost-sensitive modifications of AdaBoost (AdaCost and asymmetric boosting) are effective in detecting credit card fraud (Fan et al., 1999; Masnadi-Shirazi and Vasconcelos, 2007). AdaBoost and RL algorithms are also used to detect intrusions into computer systems and networks (Hu, Hu, and Maybank, 2008; Chen and Chen, 2009). Thus, methods of robust risk analysis, including ensemble and adaptive learning techniques, are becoming well established as tools for managing risks from intelligent adversaries.

However, the behaviors of systems of interacting intelligent agents (including software agents running their own RL algorithms, as well as humans) can be unpredictable, and low-regret policies (compared to the best that could be done with perfect information and coordination among agents on the same team) cannot necessarily be learned from data in the presence of intelligent adversaries (Yu, Mannor, and Shimkin, 2009). Moreover, while single-agent RL methods can be constrained to operate safely (avoiding acts that might cause harm) while still learning optimal control laws for engineering systems with nonlinear responses and random disturbances (e.g., in robotics or industrial process control) (Perkins and Barto, 2002), interacting adaptive controllers in multi-agent systems can settle into behavioral patterns that do not converge at all or that lead to a clearly dominated equilibrium (Busoniu, Babuska, and Schutter, 2008).

MARL algorithms are a hot research area (Dickens, Broda and Russo, 2010), with promising applications both for broad classes of decision problems, such as POMDs (Osada and Fujita, 2005), and also for practical problems such as automated trading in finance (Busoniu, Babuska, and Schutter, 2008) or detection and response to cyberterrorist distributed denial of service attacks in data networks (Xu, Sun, and Huang, 2007). However, much remains to be understood about how intelligent agents should and do coordinate, cooperate, compete, and conflict in networks and other environments before effective risk management tools can be created for the deep uncertainties created by the interaction of multiple agents.

CONCLUSIONS

For decades, the field of health, safety, and environmental risk analysis has defined itself largely in terms of providing useful answers to a few fundamental questions, such as the following: What can go wrong? How likely is it to happen? If it does happen, what are the consequences likely to be? What should we do about it? What should we say about it, how, to whom? (The first three of these questions are from Kaplan and Garrick (1981); the remaining two incorporate elements of risk management decision-making and risk communication that have been emphasized more recently.) Tools for robust risk analysis, including model ensemble, robust optimization, and adaptive learning and decision-making methods, now make it practical to refine some of these questions and to pose new ones as follows:

- Instead of (or in addition to) asking *"What can go wrong?,"* one might ask *"Is there a clearly better risk management policy than the one I am now using?"* The latter question implicitly acknowledges that not everything that might plausibly go wrong can necessarily be anticipated. What can be addressed, even with very imperfect information (e.g., in a POMDP with imprecise or unknown parameters), is whether some other policy mapping observed conditions to acts, or to probabilities of acts, would be clearly better than the current one, by any of various criteria for comparing policies in the presence of deep uncertainty (e.g., stochastic dominance, EU with imprecise probabilities, minimum EU with ambiguous probabilities, robust optimization, or measures of regret.)
- Instead of asking *"How likely is to happen?,"* one can ask *"How probable should I make each of my next possible actions?"* The probabilities of different scenarios or states or events are often unknown when decisions must be made, and they depend in part on what acts we take now and later. For example, the probability of an accident at a nuclear power plant over some time horizon depends largely on the acts and policies chosen by its operators. The probability of survival over time for a patient depends on what the physician (and, perhaps, the patient)

will do, now and later. In general, asking how likely something is to happen requires specifying what we will do, now and later. What can be answered, therefore, is not necessarily how likely different future events are, but what one will do now and what policy, mapping observations to probabilities of acts, one will use to determine what to do later. Adaptive learning policies such as SARSA and UCRL2 typically prescribe probabilities for acts to balance the two goals of maximizing rewards based on current estimates ("exploiting" what is known now) and searching for possibly better polices ("exploring" what is still uncertain).

• Instead of asking *"If it does happen, what are the consequences likely to be?,"* one can ask *"Would a different choice of policy give me lower regret (or higher EU of consequences), given my uncertainties?"* Even though the probabilities of consequences of events, given a choice of acts (and hence the immediate and delayed rewards from different act–state pairs), may be unknown or estimated only within some ranges, low-regret policies can still be developed using adaptive learning algorithms. Robust optimization can sometimes identify recommended acts even if the consequences are highly uncertain. It is therefore not necessary (and may not be possible) to predict consequences of possible future events in order to recommend low-regret or robust risk management policies. As a practical matter, decision-makers can choose policies, not events or consequences. Robust risk analysis therefore focuses on improving these choices, recognizing that event and consequence probabilities may be too uncertain to specify.

Robust risk analysis methods, including model ensemble, robust optimization, and adaptive learning and decision algorithms, shift the emphasis of the questions that define risk analysis from passive (*What might happen, and how likely is it?*) to more active (*How should I act, now and in the future?*). Risk managers are viewed not only as helping to create the future through their current decisions but also as being able to act intelligently on the basis of future information to mitigate and control risks in ways that perhaps cannot be anticipated with the more limited information available today.

Many of the future challenges for robust risk analysis will focus on changing from a single decision-maker's perspective (*What should I do?*) to a multiagent perspective (*What should we do, how might they respond, and how should we respond to their responses?*) Chapter 10 explains current approaches to answering these questions. Understanding how multiple adaptive agents collectively affect and respond to a variety of risks, from economic and financial to sociopolitical to war and terrorism, remains an outstanding challenge for the next wave of advances in robust risk analysis concepts and methods.

ACKNOWLEDGMENTS

I thank Michael Greenberg and Terje Avens for the excellent comments and suggestions that improved the exposition. Portions of this chapter have appeared in L.A. Cox, Jr., *Improving Risk Analysis* (Springer, 2013), and in L.A. Cox, Jr., Confronting deep uncertainties in risk analysis (*Risk Analysis* 2012 Oct;32(10):1607-29. doi: 10.1111/j.1539-6924. 2012.01792.x).

REFERENCES

Alagoz O, Hsu H, Schaefer AJ, Roberts MS. (2010). Markov decision processes: A tool for sequential decision making under uncertainty. Medical Decision Making 30(4):474–483.

Balaji PG, German X, Srinivasan D. (2010). Urban traffic signal control using reinforcement learning agents. Intelligent Transport Systems, IET 4(3):177–188.

Ben-Haim Y. (2001). *Information-Gap Decision Theory*. Academic Press, San Diego, CA.

Ben-Tal A, El Ghaoui L, Nemirovski A. (2009). *Robust Optimization*. Princeton University Press, Princeton, NJ.

Ben-Tal A, Bertsimas D, Brown DB (2010). A soft robust model for optimization under ambiguity. Operations Research 58(4, Part 2 of 2):1220–1234.

Bertsimas D, Brown DB, Caramanis C. (2011). Theory and applications of robust optimization. SIAM Review 53(3):464–501.

Bertsimas D, Brown DB. (2009). Constructing uncertainty sets for robust linear optimization. Operations Research 57(6):1483–1495.

Blum A, Mansour Y. (2007). From external to internal regret. The Journal of Machine Learning Research 8:1307–1324.

Bolton RJ, Hand DJ. (1999). Statistical fraud detection: A review. Statistical Science 17(3):235–255.

Bryant B, Lempert RJ. (2010). Thinking inside the box: A participatory, computer assisted approach to scenario discovery. Technology Forecasting and Social Change 77(1):34–49.

Buckley JJ. (1986). Stochastic dominance: An approach to decision making under risk. Risk Analysis 6(1):35–41.

Burton R. (2008) *On Being Certain: Believing You Are Right Even When You're Not*. St. Martin's Press, New York.

Busoniu L, Babuska R, Schutter BD. (2008). A comprehensive survey of multiagent reinforcement learning. IEEE Transactions on systems, Man, and Cybernetics-Part C: Applications and Reviews 38(2):156–172. www.sciweavers.org/publications/comprehensive-survey-multiagent-reinforcement-learning (accessed August 26, 2014).

Cai C, Liao X, Cari L. (2009). Learning to explore and exploit in POMDPs. Advances in Neural Information Processing Systems 22:198–206. http://people.ee.duke.edu/~lcarin/LearnE2_NIPS09_22_FINAL.pdf (accessed August 26, 2014).

Carpenter TE, O'Brien JM, Hagerman AD, McCarl BA. (2011). Epidemic and economic impacts of delayed detection of foot-and-mouth disease: A case study of a simulated outbreak in California. Journal of Veterinary Diagnostic Investigation 23(1):26–33.

Cesa-Bianchi N, Lugosi G. (2006). *Prediction, learning, and games*. Cambridge University Press, Cambridge.

Chades I, Bouteiller B. (2005). Solving multiagent Markov decision processes: A forest management example. In MODSIM 2005 International Congress on Modelling and Simulation, Hampton, VA on October 13–15, 2010. A. Zerger and R.M. Argent Eds. Modelling and Simulation Society of Australia and New Zealand.

Chen Y. Chen Y. (2009). Combining incremental Hidden Markov Model and Adaboost algorithm for anomaly intrusion detection. In Proceedings of the ACM SIGKDD Workshop on Cybersecurity and intelligence informatics (Paris, France, June 28–28, 2009). H. Chen, M. Dacier, M. Moens, G. Paass, and C.C. Yang, Eds. CSI-KDD '09. ACM, New York, 3–9.

Churchman, CW, (1967). Wicked problems. Management Science 14(4):B141–B142.

Condorcet NC de. (1785). *Essai sur l'Application de l'Analyse a la Probabilite des Decisions Rendues a la Pluralite des voix*. Imprimerie Royale, Paris.

Cortés EA, Gámez M, Rubio NG. (2007). Multiclass corporate failure prediction by Adaboost.M1. International Advances in Economic Research 13(3):301–312.

Dalamagkidis D, Kolokotsa D, Kalaitzakis K, Stavrakakis GS. (2007). Reinforcement learning for energy conservation and comfort in buildings. Building and Environment 42:2686–2698. http://www.tuc.gr/fileadmin/users_data/elci/Kalaitzakis/J.38.pdf (accessed August 27, 2014).

Das TK, Savachkin AA, Zhu Y. (2007) A large scale simulation model of pandemic influenza outbreaks for development of dynamic mitigation strategies. IIE Transactions 40(9):893–905. http://www.eng.usf.edu/~das/papers/das_r1.pdf (accessed August 27, 2014).

Dickens L, Broda K, Russo A. (2010). The dynamics of multi-agent reinforcement learning. In Frontiers in Artificial Intelligence and Applications Volume 215, Proceedings of the 2010 conference on ECAI 2010: 19th European Conference on Artificial Intelligence. H. Coelho, R. Studer, and M. Wooldridge, Eds. http://www.doc.ic.ac.uk/~lwd03/ecai2010.pdf (accessed August 27, 2014).

Ernst D, Stan G-B, Gongalves J, Wehenkel L. (2006). Clinical data based optimal STI strategies for HIV: A reinforcement learning approach. In 45th IEEE Conference on Decision and Control, December 13–15, San Diego, CA, 667–672. http://www.montefiore.ulg.ac.be/~stan/CDC_2006.pdf (accessed August 27, 2014).

Fan W, Stolfo S, Zhang J, Chan P. (1999). Adacost: Misclassification cost-sensitive boosting. In Proceedings of the 16th International Conference on Machine Learning, Bled, Slovenia, Morgan Kaufmann Publishers Inc. San Francisco, CA, 97–105. http://dl.acm.org/citation.cfm?id=657651 (accessed December 4, 2014).

Fiege J, McCurdy B, Potrebko P, Champion H, Cull A. (2011). PARETO: A novel evolutionary optimization approach to multiobjective IMRTs planning. Medical Physics 38(9):5217–5229.

Forsell N, Garcia F, Sabbadin R. (2009). Reinforcement learning for spatial processes. In World IMACS / MODSIM Congress, July 13–17, 2009 Cairns, Australia, http://www.mssanz.org.au/modsim09/C1/forsell.pdf (accessed August 27, 2014).

Fredriksson A, Forsgren A, Hårdemark B. 2011. Minimax optimization for handling range and setup uncertainties in proton therapy. Medical Physics 38(3): 1672–1684.

Ge L, Mourits MC, Kristensen AR, Huirne RB. (2010). A modelling approach to support dynamic decision-making in the control of FMD epidemics. Preventive Veterinary Medicine 95(3–4):167–174.

Geibel P, Wysotzk F. (2005). Risk-sensitive reinforcement learning applied to control under constraint. Journal of Artificial Intelligence Research 24:81–108.

Gilboa I, Schmeidler D. (1989). Maxmin expected utility with a non-unique prior. Journal of Mathematical Economics 18:141–153.

Green CS, Benson C, Kersten D, Schrater P. (2010). Alterations in choice behavior by manipulations of world model. Proceedings of the National Academy of Sciences of the United States of America 107(37):16401–16406.

Gregoire PL, Desjardins C, Laumonier J, Chaib-draa B. (2007). Urban traffic control based on learning agents. In Intelligent Transportation Systems Conference, ITSC 2007, IEEE, Seattle, WA, 916–921. ISBN: 978-1-4244-1396-6. doi:10.1109/ITSC.2007.4357719.

Hansen LP, Sargent TJ. (2001). Robust control and model uncertainty. The American Economic Review, 91:60–66.

Hansen LP, Sargent TJ. (2008). *Robustness*. Princeton University Press, Princeton, NJ.

Harford T. (2011). *Adapt: Why Success Always Starts with Failure*. Farra, Straus and Giroux, New York.

Hauskrecht M, Fraser H. (2000). Planning treatment of ischemic heart disease with partially observable Markov decision processes. Artificial Intelligence in Medicine 18(3):221–244.

Hazen E, Seshadhri C. (2007). Efficient learning algorithms for changing environments. In ICML '09 Proceedings of the 26th Annual International Conference on Machine Learning, New York. http://ie.technion.ac.il/~ehazan/papers/adap-icml 2009.pdf (accessed August 27, 2014).

Hoeting JA, Madigan D, Raftery AE, Volinsky CT. (1999). Bayesian Model Averaging: A Tutorial. Statistical Science 14(4):382–401 http://mpdc.mae.cornell.edu/Courses/UQ/2676803.pdf

Hrdlicka J, Klema J. (2011). Schizophrenia prediction with the adaboost algorithm. Studies in Health Technology and Informatics 169:574–578.

Hu W, Hu W, Maybank S. (2008). AdaBoost-based algorithm for network intrusion detection. IEEE Transactions on Systems, Man, and Cybernetics. Part B, Cybernetics 38(2):577–583.

Hutter M, Poland J. (2005). Adaptive online prediction by following the perturbed leader. Journal of Machine Learning Research 6:639–660. http://jmlr.csail.mit.edu/papers/volume6/hutter05a/hutter05a.pdf (accessed August 27, 2014).

Inaniwa T, Kanematsu N, Furukawa T, Hasegawa A. (2011). A robust algorithm of intensity modulated proton therapy for critical tissue sparing and target coverage. Physics in Medicine and Biology 56(15):4749–4770.

Itoh H, Nakamura K. (2007). Partially observable Markov decision processes with imprecise parameters. Artificial Intelligence 171(8–9):453–490.

Izadi MT, Buckeridge DL. (2007). Optimizing anthrax outbreak detection using reinforcement learning. In IAAI'07 Proceedings of the 19th national conference on Innovative applications of artificial intelligence—Volume 2. AAAI Press, Vancouver. http://www.aaai.org/Papers/AAAI/2007/AAAI07-286.pdf (accessed August 27, 2014).

Jaksch T, Ortner R, Auer P. (2010). Near-optimal regret bounds for reinforcement learning. Journal of Machine Learning Research 11:1563–1600.

Jung J, Liu CC, Tanimoto S, Vittal V. (2002). Adaptation in load shedding under vulnerable operating conditions. IEEE Transactions on Power Systems 17:1199–1205.

Kahnt T, Park SQ, Cohen MX, Beck A, Heinz A, Wrase J. (2009). Dorsal striatal-midbrain connectivity in humans predicts how reinforcements are used to guide decisions. Journal of Cognitive Neuroscience 21(7):1332–45.

Kaplan S, Garrick BJ, (1981). On the quantitative definition of risk. Risk Analysis 1(1):11–27. http://josiah.berkeley.edu/2007Fall/NE275/CourseReader/3.pdf (accessed August 27, 2014).

Koop G, Tole L. (2004). Measuring the health effects of air pollution: To what extent can we really say that people are dying from bad air? Journal of Environmental Economics and Management 47:30–54. http://citeseerx.ist.psu.edu/viewdoc/summary?doi=10.1.1.164.6048 (accessed August 27, 2014).

Kuyer L, Whiteson S, Bakker B, Vlassis N. (2008). Multiagent reinforcement learning for urban traffic control using coordination graphs. In ECML 2008: Proceedings of the Nineteenth European Conference on Machine Learning, 656–671.

Laeven R, Stadje MA. (2011). Entropy coherent and entropy convex measures of risk. Tilburg University CentER Discussion Paper 2011–031. http://arno.uvt.nl/show.cgi?fid=114115 (accessed August 27, 2014).

Lee EK, Chen CH, Pietz F, Benecke B. (2010). Disease propagation analysis and mitigation strategies for effective mass dispensing. AMIA Annual Symposium Proceedings 2010:427–431.

Lempert RJ, Collins MT. (2007). Managing the risk of uncertain threshold response: Comparison of robust, optimum, and precautionary approaches. Risk Analysis 27(4):1009–1026.

Lempert R, Kalra N. (2008). *Managing Climate Risks in Developing Countries with Robust Decision Making*. World Resources Report, Washington, DC. http://www.worldresourcesreport.org/files/wrr/papers/wrr_lempert_and_kalra_uncertainty.pdf (accessed August 27, 2014).

Lizotte DJ, Gunter L, Laber E, Murphy SA. (2008) *Missing data and uncertainty in batch reinforcement learning*, NIPS-08 Workshop on Model Uncertainty and Risk in RL. http://www.cs.uwaterloo.ca/~ppoupart/nips08-workshop/nips08-workshop-schedule.html (accessed August 27, 2014).

Lu F, Boritz JE, Covvey HD. (2006). Adaptive fraud detection using Benford's law. In Advances in Artificial Intelligence: 19th Conference of the Canadian Society for Computational Studies of Intelligence, Quebec city. http://bit.csc.lsu.edu/~jianhua/petrov.pdf (accessed August 27, 2014).

Maccheroni F, Marinacci M, Rustichini A. (2006). Ambiguity aversion, robustness, and the variational representation of preferences. Econometrica 74:1447–1498.

Makridakis S, Hibon M. (2000). The M3-Competition: Results, conclusions and implications. International Journal of Forecasting 16:451–476. http://www.forecastingprinciples.com/files/pdf/Makridakia-The%20M3%20Competition.pdf (accessed August 27, 2014).

Masnadi-Shirazi H, Vasconcelos N. (2007). Asymmetric boosting. In Proceedings of the 24th International Conference on Machine Learning, Oregon State University, Corvallis, OR, June 20–24, 2007, ACM, New York, 609–619. http://oregonstate.edu/conferences/event/icml2007/ (accessed December 4, 2014).

McDonald-Madden E, Chadès I, McCarthy MA, Linkie M, Possingham HP. (2011). Allocating conservation resources between areas where persistence of a species is uncertain. Ecological Applications 21(3):844–58.

Molinaro AM, Simon R, Pfeiffer RM. (2005). Prediction error estimation: A comparison of resampling methods. Bioinformatics 21(15):3301–3307.

Morra JH, Tu Z, Apostolova LG, Green AE, Toga AW, Thompson PM. (2010). Comparison of AdaBoost and support vector machines for detecting Alzheimer's disease through automated hippocampal segmentation. IEEE Transactions on Medical Imaging 29(1):30–43.

Ni Y, Liu Z-Q. (2008). Bounded-parameter partially observable Markov decision Processes. In Proceedings of the Eighteenth International Conference on Automated Planning and Scheduling, Sydney, September 14–18, 2008 (Jussi Rintanen, Bernhard Nebel, J. Christopher Beck, and Eric Hansen, eds.). The AAAI Press, Menlo Park. http://www.aaai.org/Library/ICAPS/icaps08contents.php (accessed December 4, 2014).

Niua B, Jinb Y, Lua WC, Li GZ. (2009). Predicting toxic action mechanisms of phenols using AdaBoost learner. Chemometrics and Intelligent Laboratory Systems 96(1):43–48.

Osada H, Fujita S. (2005). CHQ: A multi-agent reinforcement learning scheme for partially observable Markov decision processes. IEICE—Transactions on Information and Systems E88-D(5):1004–1011.

Perkins TJ, Barto AG. (2002). Lyapunov design for safe reinforcement learning. Journal of Machine Learning Research 3:803–883. http://jmlr.csail.mit.edu/papers/volume3/perkins02a/perkins02a.pdf (accessed August 27, 2014).

Regan K, Boutilier C. (2008). Regret-based reward elicitation for Markov decision processes. NIPS-08 Workshop on Model Uncertainty and Risk in RL. http://www.cs.uwaterloo.ca/~ppoupart/nips08-workshop/nips08-workshop-schedule.html (accessed August 27, 2014).

Rittel H, Webber M. (1973). Dilemmas in a general theory of planning. Policy Sciences (4):155–169. [Reprinted in N. Cross (ed.), *Developments in Design Methodology*, John Wiley & Sons, Ltd, Chichester, 1984, 135–144.] http://www.uctc.net/mwebber/Rittel+Webber+Dilemmas+General_Theory_of_Planning.pdf (accessed August 27, 2014).

Ross S, Pineau J, Chaib-draa B, Kreitmann P. (2011). POMDPs: A new perspective on the explore-exploit tradeoff in partially observable domains. Journal of Machine Learning Research (12):1729–1770.

Sabbadin R, Spring D, Bergonnier E. A Reinforcement-Learning Application to Biodiversity Conservation in Costa-Rican Forest. In 17th International Congress on Modelling and Simulation (MODSIM'07). http://www.mssanz.org.au/MODSIM07/papers/41_s34/AReinforcement_s34_Sabbadin_.pdf (accessed December 4, 2014).

Savio A, García-Sebastián M, Graña M, Villanúa J. (2009). Results of an Adaboost approach on Alzheimer's disease detection on MRI. Bioinspired Applications in Artificial and Natural Computation Lecture Notes in Computer Science 5602/2009: 114–123. www.ehu.es/ccwintco/uploads/1/11/GarciaSebastianSavio-VBM_SPM_SVM-IWINAC2009_v2.pdf (accessed August 27, 2014).

Schaefer AJ, Bailey MD, Shechter SM, Roberts MS. (2004). Modeling medical treatment using Markov decision processes. In *Handbook of Operations Research/Management Science Applications in Health Care*. Kluwer Academic Publishers, Boston, MA, 593–612. http://www.ie.pitt.edu/~schaefer/Papers/MDPMedTreatment.pdf (accessed August 27, 2014).

Schönberg T, Daw ND, Joel D, O'Doherty JP. (2007). Reinforcement learning signals in the human striatum distinguish learners from nonlearners during reward-based decision making. The Journal of Neuroscience 27(47):12860–12867.

Srinivasan J, Gadgil S. (2002). Asian brown cloud-fact and fantasy. Current Science 83:586–592.

Su Q, Lu W, Niu B, Liu X. (2011). Classification of the toxicity of some organic compounds to tadpoles (Rana Temporaria) through integrating multiple classifiers. Molecular Informatics 30(8):672–675.

Sutton RS, Barto AG. (2005). *Reinforcement Learning: An Introduction*. MIT Press, Cambridge, MA. http://rlai.cs.ualberta.ca/~sutton/book/ebook/the-book.html (accessed August 27, 2014).

Svetnik V, Wang T, Tong C, Liaw A, Sheridan RP, Song Q. (2005). Boosting: An ensemble learning tool for compound classification and QSAR modeling. Journal of Chemical Information and Modeling 45(3):786–799.

Szepesvari C. (2010). *Reinforcement Learning Algorithms*. Morgan & Claypool Publishers. http://www.morganclaypool.com/ (accessed December 4, 2014).

Tan C, Chen H, Xia C. (2009). Early prediction of lung cancer based on the combination of trace element analysis in urine and an Adaboost algorithm. Journal of Pharmaceutical and Biomedical Analysis 49(3):746–752.

Walker WE, Marchau VAWJ, Swanson D. (2010). Addressing deep uncertainty using adaptive policies introduction to section 2. Technological forecasting and social change 77(6):917–923.

Waltman L, van Eck NJ. (2009). Robust evolutionary algorithm design for socio-economic simulation: Some comments. Computational Economics 33:103–105. http://repub.eur.nl/res/pub/18660/RobustEvolutionary_2008.pdf (accessed August 27, 2014).

Wang Y, Xie Q, Ammari A. (2011). Deriving a near-optimal power management policy using model-free reinforcement learning and Bayesian classification. In DAC '11 Proceedings of the 48th Design Automation Conference, San Diego, CA, USA, June 5–10, 2011, ACM, New York. http://dl.acm.org/citation.cfm?id=2024724 (accessed December 4, 2014).

Weick KE, Sutcliffe KM. (2007). *Managing the Unexpected: Resilient Performance in an Age of Uncertainty*, 2nd Ed. John Wiley & Sons, Inc., Hoboken, NJ.

Yang CC, Zeng DD, Chau M, Chang K, Yang Q, Cheng X, Wang J, Wang F-Y, Chen H. (2007). Intelligence and Security Informatics, Pacific Asia Workshop, PAISI 2007, Chengdu, China, April 11–12, 2007, Proceedings. Lecture Notes in Computer Science 4430, Springer. ISBN 978-3-540-71548-1. http://dblp.uni-trier.de/db/conf/paisi/paisi2007.html (accessed December 4, 2014).

Yousefpour R, Hanewinkel M. (2009). Modelling of forest conversion planning with an adaptive simulation-optimization approach and simultaneous consideration of the values of timber, carbon and biodiversity. Ecological Economics 68(6):1711–1722.

Ye D, Zhang M, Sutato D. (2011). A hybrid multiagent framework with Q-learning for power grid systems restoration. IEEE Transactions on Power Systems 26(4): 2434–2441.

Yu JY, Mannor S, Shimkin N. (2009). Markov decision processes with arbitrary reward processes. Mathematics of Operations Research 34(3):737–757.

Zhao Y, Kosorok MR, Zeng D. (2009). Reinforcement learning design for cancer clinical trials. Statistics in Medicine 28(26):3294–3315.

Zhou B, Chan KW, Yu T. (2011). Q-Learning approach for hierarchical AGC scheme of interconnected power grids. The Proceedings of International Conference on Smart Grid and Clean Energy Technologies. Energy Procedia 12:43–52.

Zhou L, Kin Keung Lai. (2009). Adaboosting neural networks for credit scoring. Advances in Intelligent and Soft Computing 56/2009:875–884. doi:10.1007/978-3-642-01216-7_93.

CHAPTER 8

Medical Decision-Making: An Application to Sugar-Sweetened Beverages

Andrea C. Hupman[1] and Ali E. Abbas[2]

[1] Department of Industrial and Enterprise Systems Engineering, College of Engineering, University of Illinois at Urbana-Champaign, Urbana-Champaign, IL, USA

[2] Industrial and Systems Engineering, Center for Risk and Economic Analysis of Terrorism Events, Sol Price School of Public Policy, University of Southern California, Los Angeles, CA, USA

This is the first of three chapters that focus on real-world applications of decision and risk analysis. In this chapter, the main technical methods used are multiattribute utility theory and decision tree analysis. These are classical decision analysis topics that would have been familiar to practitioners 30 years ago, but in this chapter they are applied to the challenging domain of valuing risks to life and health and trading them off against the pleasure derived from consumption. Chapters 9 and 10 consider applications of decision and risk analysis to building and defending resilient infrastructures and defending against attacks by terrorists or other intelligent adversaries. All three chapters emphasize the practical applicability of decision and risk analysis methods to important decisions that are

Breakthroughs in Decision Science and Risk Analysis, First Edition.
Edited by Louis Anthony Cox, Jr.
© 2015 John Wiley & Sons, Inc. Published 2015 by John Wiley & Sons, Inc.

often made without the benefit of such formal methods and that can potentially be made better with them.

INTRODUCTION

What comes to mind when you hear the term "medical decision-making?" Common answers include hospitals, doctors, shots, medical tests, and treatments—things associated with illness. But medical decision-making encompasses topics broader than treating illness; it includes maintaining wellness. The decisions we make while we are healthy have a tremendous impact on long-term health and mortality. Analysis has shown that individuals' own decisions are the leading cause of death in the United States (Keeney, 2008). This chapter discusses the decision analytic foundation of medical decision-making and how it is applied to decisions that affect life and health.

A general definition of medical decision-making is any decision situation that affects the provision of medical care or the medical well-being of individuals. Medical decision-making can refer to medical diagnostics, medical treatment selection, patient scheduling, public health decisions, as well as the impact of personal decisions on health, the focus of this chapter. Two recurrent themes across the applications of medical decision-making are the importance of decision-making on health outcomes and the importance of decision analytic principles. This chapter highlights these themes throughout its treatment of decisions on life and health.

The significance of the impact our decisions have on health cannot be overstated. The probability of each possible health outcome faced by an individual is determined by the sequence of decisions leading up to that outcome. In medical decision-making, one potential outcome is death. Consider the sequence of decisions involved for an individual who finds a lump in her breast. The decisions to perform a self-breast exam and to seek medical attention affect the severity of the condition when medical guidance is first sought. If the lump is determined to be cancerous, the probability of survival and the treatment prospects change based on the stage of the cancer. Thus, the sequence of decisions involved affect whether the individual will live or die, clearly illustrating the importance of medical decisions.

To analyze medical decisions, we rely on the elements of decision quality: a committed decision-maker, sound reasoning (the logic by which we choose), a set of feasible decision alternatives, a decision frame that helps us focus on the right problem, good information, and clear values and trade-offs. Several of these components are largely dictated by the situation that will determine, for example, what information is relevant and what alternatives are possible. The element of specifying values, however, depends entirely on the decision-maker. In medical decisions, it is important to know how the decision-maker values his/her life and his/her monetary consumption. Such questions can be difficult and uncomfortable for some people to contemplate because although people are exposed to risks of death every day, they are often not aware of them. Medical decision-making requires people consider their own mortality, a necessity that often makes medical decision-making difficult.

As discussed in each of the previous chapters, the presence of uncertainty often adds difficulty to decision-making. Even when the decision-maker has all the relevant information and is clear about his/her values, he/she cannot be guaranteed a good outcome when uncertainty is present. Consider the individual who exercised regularly and maintains a healthy diet. In spite of these healthy decisions, there is no guarantee that he will not develop a grave medical condition such as cancer. This presence of uncertainty only compounds the difficulty of medical decision-making where the potential outcomes may include disability or death.

Our purpose in this chapter is to present medical decision-making from a modern decision analytic perspective. All the elements of decision quality are needed in medical decision-making, but we focus on the element of preferences. It can be assumed that better health is preferred to poorer health and that longer life in a good health state is preferred to shorter life in that state. But the relative preferences for different health prospects depend on the decision-maker. This chapter explains how medical preferences can be modeled with a multiattribute utility function and how these preferences can be used within a decision tree framework to analyze medical decisions.

In addition to preferences, the decision tree framework incorporates the other elements of decision quality. The construction of the decision tree is dictated by the decision frame or perspective. The decision alternatives are included in the form of choice nodes. Relevant information is included in the form of the probabilities at chance nodes. Information may also be incorporated in the tree through the generation of additional alternatives and the better characterization of potential outcomes. With the decision alternatives and their respective potential outcomes described, a multiattribute utility function is used to model the decision-maker's preferences under conditions of uncertainty. The expected utility associated with each alternative can then be determined. Using the logic of decision analysis, the preferred decision alternative can be determined.

To facilitate the explanation of decision analytic concepts in decisions affecting life and health, we present an illustrative example on the consumption of sugar-sweetened beverages (SSB). We explain the analysis framework and compile the data necessary for analysis. The selection of this particular example is motivated by research that has associated SSB consumption with an increased risk of type 2 diabetes (Schulze et al., 2004) and by the significant impact of type 2 diabetes on society. Over 8% of the US population is affected by diabetes, which is the leading cause of blindness, nontraumatic lower limb amputation, and kidney failure. It is the seventh leading cause of death in the Unites States (Centers for Disease Control and Prevention (CDC), 2011b).

While our discussion of medical decision-making is focused around the decision to consume SSB, the same concepts are fundamental to the analysis of any medical decision, and the approach is applicable to a variety of medical decision-making situations.

The chapter begins with a brief treatment of medical ethics relevant to the discussion of medical decision analysis. We then explain a multi-attribute utility model of life and consumption before presenting the decision analytic framework. We discuss how the analysis for an individual decision can be used to derive insights of the implications for society at large. Finally, we discuss generalizations of the framework and alternative approaches to medical decision-making.

MEDICAL ETHICS AND AUTONOMY

Discussion of medical ethics is based on four guiding principles: autonomy, nonmaleficence, beneficence, and justice (Beauchamp and Childress, 2001). Within the context of decision-making, the principle of autonomy is particularly relevant. Autonomy refers to an individual's right to self-determination, a right that indicates an individual has the right to make the decisions that will affect his/her health well-being.

Decision-making involves specifying values. Each person has his/her own independent values. The diversity of individual values can lead to the existence of controversial choices (Savulescu, 2007). In this chapter, we assume each individual retains autonomy over his/her own decisions. We further assume each individual is mentally competent to fully understand the decision at hand and potential consequences. Any value we calculate is specific to each individual and indeed can only be paid by that individual as it is determined by changes to life expectancy.

MULTIATTRIBUTE UTILITY FOR PREFERENCES OF LIFE AND CONSUMPTION UNDER UNCERTAINTY

Medical decision-making involves making decisions when the outcomes are uncertain. The tenants of decision analysis are used in valuing preferences when uncertainty is present (Howard and Abbas, 2015). The rules of decision analysis are the following:

1. The decision-maker can specify the probability of each possible prospect following the decision.
2. The decision-maker can arrange the decision prospects in order of preference.
3. In an uncertain deal with a probability p of obtaining the most preferred prospect and a probability $1-p$ of the least preferred prospect, the decision-maker can specify a probability p that makes him indifferent between this deal and a decision prospect.
4. The decision-maker is indifferent between obtaining any decision prospect and its respective uncertain deal as described in (3) so that all decision alternatives may be substituted with binary deals with the same two prospects.
5. The decision-maker chooses the alternative that is substituted for the uncertain deal with the greatest probability of the most preferred prospect.

Maximizing expected utility is consistent with following these five rules. Specifying a utility function to describe preferences under uncertainty facilitates decision-making. Instead of specifying indifference probabilities in uncertain deals for each prospect, the decision-maker specifies the attributes of each prospect that affect his/her preferences for that prospect and specifies his/her preferences for the attributes. For example, in medical decision-making, length of life might be an

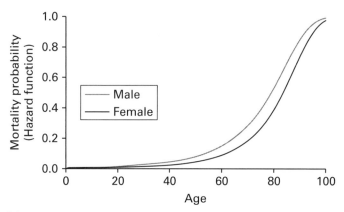

Figure 8.1 Population mortality by age and gender (data from the US Census; Arias, 2010).

important attribute to the decision-maker. When multiple attributes are present, a multiattribute utility function is used to describe the decision-maker's preferences under uncertainty.

One useful multiattribute utility model describes preferences over length of life and consumption through trade-offs between the two attributes (Howard, 1984). The parameters necessary to implement this model are the individual's life expectancy $(\bar{l})$, risk tolerance (ρ), constant annual consumption (c), and trade-off parameter (η). These parameters are explained in more detail in the following paragraphs.

The *life expectancy* $(\bar{l})$ for a specific population can be obtained from government agencies that track mortality data. In the United States, mortality data is available from the Department of Health and Human Services (DHHS) and is illustrated in Figure 8.1 (Arias, 2010).

Constant annual consumption (c) is also determined by the decision-maker and is determined based on his/her estimation of his/her future prospects. The decision-maker is asked to think about his/her future consumption for the rest of his/her life that is above bare survival and how it may change with time. Next, the decision-maker is asked to consider receiving a constant amount each year in place of his/her variable consumption. The decision-maker is asked to determine what constant consumption makes him just indifferent to receiving this constant amount or receiving his/her uncertain future consumption. Converting the uncertain variable consumption into a constant annual consumption, denoted c, will simplify the analysis. Note the distinction that consumption is above what is required for survival. If consumption is defined differently, the model will need to take this into account.

The *trade-off parameter* η governs the decision-maker's relative preference for years lived versus consumption. It is the percentage decrease in constant annual consumption the decision-maker would accept for his/her life to be 1% longer:

$$\eta = -\frac{\Delta c / c}{\Delta \bar{l} / l}.$$
(8.1)

The parameter η is assessed for small changes in life expectancy. As life expectancy decreases, the decision-maker may refuse to trade any life expectancy for consumption. Thus, this model is best suited to small percentage changes in life and consumption. The parameter η need not be constant but could be a function of life expectancy.

Risk tolerance (ρ) describes the decision-maker's attitude toward risk and uncertainty. It must be assessed from the decision-maker because it is specific to his/her preferences under conditions of uncertainty. A low risk tolerance indicates an unwillingness to accept uncertain deals or the necessity to be paid a high premium in order to accept uncertainty.

To assess risk tolerance, the decision-maker is asked to consider an uncertain deal in which there is a probability p of doubling his/her constant annual consumption for the remainder of his/her life when he is guaranteed to live his/her expected lifetime. With probability $(1-p)$, however, his/her constant annual consumption will be halved. The decision-maker is asked to determine the probability p that would make him just indifferent between his/her current constant annual consumption and the uncertain deal. The risk tolerance is the value that makes the decision-maker's annual certain equivalent for consumption equal to his/her constant annual consumption.

With these attributes and parameters defined, the utility of consumption and life length is given as

$$U(c,l) = 1 - \exp\left[-\frac{c}{\rho}\left(\frac{l}{\bar{l}}\right)^{\eta}\right].$$
(8.2)

It is an exponential utility function over the value determined by the trade-off between consumption and length of life, $c\left(l / \bar{l}\right)^{\eta}$. In general, any utility functional form could be used. In general, any utility functional form could be used that matches the decision maker's preferences. Note that (8.2) has several properties that are well suited to a medical context. It exhibits attribute dominance (Abbas and Howard, 2005), meaning that if either attribute is zero, then the entire utility function is zero. In addition, (8.2) does not exhibit utility independence. We leave it to the reader to consider why utility independence need not hold in the context of most medical decisions.

This multiattribute utility function can be used to calculate the value of medical decisions. We next discuss a decision analytic framework in which this model can be used to value the consumption of SSB.

ANALYSIS FORMULATION

The analysis of a decision begins with determining the decision frame or perspective. The frame will determine how the decision is defined. For example, an individual wishing to lose weight may consider surgical options and frame the decision as which type of weight loss surgery to get. Another decision frame would lead the decision-maker to decide between which type of diet to follow or to decide which exercise plan to adopt. The frame determines the remainder of the analysis.

Once the frame is determined, the decision tree can be constructed. The tree will model the probability of each potential outcome of each decision alternative. Each outcome is described in terms of the important attributes defined by the decision-maker. The preference for each outcome is described using an appropriate multiattribute utility function.

This section presents an illustrative example to explain the construction of a decision tree in a medical decision-making application. The example decision is on the consumption of SSB, which has been shown to increase the risk of developing type 2 diabetes (Schulze et al., 2004). Diabetes, in turn, increases the mortality risk by a factor of two (CDC, 2011b).

We first describe the decision tree framework and the appropriate calculations. We then discuss special considerations and limitations of such analyses. Finally, we present an analysis of the case example for determining the value of a medical decision.

The Decision Tree Framework

A decision tree consists of decision nodes, represented by squares, and uncertainty nodes represented by circles. For decisions that affect probabilities over several years, multiple uncertainty nodes may be used

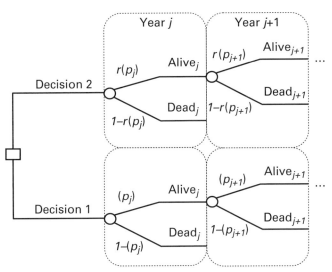

Figure 8.2 General decision tree framework for a decision that affects the risk of death in future life years.

sequentially. A general model of a decision that affects the probability of death for the remainder of one's life is illustrated in Figure 8.2 where r represents the effect of the decision on the survival probability of each year of life, p_j.

The general decision tree of Figure 8.2 can be tailored for use in a variety of decision situations. For example, consider decisions that affect risk of disease such as the risk of type 2 diabetes. We model the decision to consume SSB and its effect on the risk of type 2 diabetes in Figure 8.3. The notation is summarized in Table 8.1. The individual faces a decision, and the resulting risks are modeled sequentially by year, with each year denoted by the subscript j.

In each year following the decision, the individual faces a risk of death, regardless of the decision made. The probability of remaining alive at the end of year j given the individual is alive at the beginning of the year is denoted p. The probability of death is $1 - p_j$.

In each year the decision-maker remains alive, he/she faces a risk of developing diabetes in that year. The probability of being diagnosed with type 2 diabetes in a given year, conditional on being alive in that year, is denoted d. The relative risk of type 2 diabetes associated with the decision is a constant multiplier denoted r. Once he/she develops diabetes, his/her mortality probabilities increase; this change is not

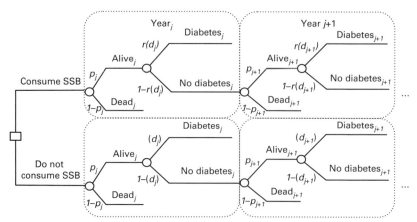

Figure 8.3 Decision tree framework for determining the value of a personal decision.

Table 8.1 Notation used in the model

Notation	Definition
p_j	The probability of being alive at the end of year j for the nondiabetic individual, conditioned on being alive at the beginning of the year
d_j	The probability of developing diabetes in year j
r	The relative risk of type 2 diabetes associated with the decision

illustrated in Figure 8.3. If he/she does not develop diabetes, he/she continues each year as before, with a probability of being alive and a probability of developing diabetes.

Calculations

The probability of a particular outcome is determined by following the path on the tree to that outcome. For example, the probability of being alive without a diagnosis of diabetes is calculated

$$p(\text{Alive}_j, \text{No Diabetes}_j) = \prod_{i=\text{Current Age}}^{j} p_i(1 - d_i). \qquad (8.3)$$

The probability of dying at age j without having diabetes is

$$p(\text{Die}_j, \text{No Diabetes}_j) = p(\text{Alive}_{j-1}, \text{No Diabetes}_{j-1})(1 - p_j). \qquad (8.4)$$

The probability of being diagnosed with diabetes at a future age requires that the individual is alive and has not previously been diagnosed. It is calculated

$$p(\text{Alive}_j, \text{Diabetes}_j) = p(\text{Alive}_{j-1}, \text{No Diabetes}_{j-1}) p_j d_j. \quad (8.5)$$

The probability of being diagnosed with diabetes at age i and dying at age j is

$$p(\text{Die}_j, \text{Diabetes}_i) = p(\text{Alive}_i, \text{Diabetes}_i)\left(1 - p_{j,\text{Diabetes}}\right) \prod_{k=i+1}^{j-1} p_{k,\text{Diabetes}},$$

$$(8.6)$$

where $p_{j,\text{Diabetes}}$ is the probability of being alive at the end of year j for the individual with diabetes, conditioned on being alive at the beginning of the year.

In some medical decisions, the possible outcomes are the same regardless of the decision made, but the probability distribution over the outcomes changes. This is the case for consuming SSB where the difference in the outcomes of the decision tree lies solely in the different risks of diabetes. The probability nodes in the decision alternative that does not change the individual's risk of diabetes are the baseline risks of the population, d_j. The probability nodes in the decision alternative with an increased risk of type 2 diabetes are modified by a multiplier, $r(d_j)$. Because the relative risks are determined through experimental studies specific to each probability d_j, the probability at any given node in the tree is prevented from having a probability greater than 1. A benefit of this approach is that the model is flexible to analyze any decision for which the increased risk of a medical condition is known or can be estimated.

Analysis of such decisions enables the determination of the value of a decision in terms of its effect on lost life. The decision-maker's value per SSB is calculated using an appropriate multiattribute utility model where one of the attributes is consumption. Each terminating branch of the tree represents one possibility and has a utility value. We find the probability of that outcome occurring and then can find the expected utility for that decision alternative in the tree. Optimization software is used to find the consumption change needed for the expected utility to be equal in each decision alternative in Figure 8.3. This change in consumption is used to determine the value of each SSB.

Special Considerations and Limitations

In the application of any model, it is important to understand the underlying assumptions and potential limitations of the model. In addition, the model will only be as accurate as the data populating it. We discuss some limitations and special considerations in the application of data to this decision framework and specifically in the case of analyzing SSB consumption.

It must be possible to characterize the risks associated with the decision to conduct analysis. In the analysis of SSB consumption, we use constant risk multipliers as reported in the literature for the risk of type 2 diabetes associated with SSB consumption (Schulze et al., 2004). These risks are reported as constant. However, it may be possible that they change over time. Any such changes over time will not be reflected in the model. These reported risks have been adjusted for numerous factors including age, physical activity, and family history, but any limitations in these reported risks will be reflected in the model. For example, the data available used only women as test subjects; similar data for males is not available. Thus, any analysis for men must make assumptions on the applicability of data derived from women.

The baseline population risks for diabetes used are those reported in the literature (Narayan et al., 2003). We do not specify other factors that affect one's risk of type 2 diabetes. For example, family history is a known risk factor. To include other such factors, a function of the baseline population risks may be used, $f(p_j)$. However, in this case, the data on the relative risk of the decision specific to individuals with given additional risk factors may not be available. Thus, assumptions still need to be made regarding the relative risk that will affect the analysis results.

The mortality for individuals with diabetes is the reported population average. If an individual manages his/her diabetes much better than the average person, his/her mortality probabilities will likely be lower than the population average. These considerations highlight the limitations of the framework.

Another consideration in the framework is the difference in how the mortality probability obtained from the US DHHS is calculated and in how the relative risk of mortality for individuals with diabetes is calculated. The age-specific mortality probabilities from the US

DHHS include individuals with and without diabetes. The relative risk of mortality for those with diabetes is relative to individuals without diabetes. To avoid overestimating the effect of diabetes on mortality, we can calculate separate mortality probabilities by diabetes state.

We introduce some new notation for these calculations. We let the mortality probabilities obtained for the population be denoted $p_{j,\text{Population}}$. We let N_j and $\bar{N}_j$ denote the total number of individuals with and without diabetes alive at the beginning of the year. This data is available from the US National Health Interview Survey (CDC, 2011a). We let n_j and $\bar{n}_j$ denote the number of individuals with and without diabetes, respectively, who die at age j.

To determine the separate mortality probabilities by diabetes status, we will derive two expressions that allow us to solve for the unknown quantities n_j and $\bar{n}_j$. The probabilities we are looking for will then be calculated as the number of individuals with or without diabetes who die in year j divided by those alive at the beginning of the year:

$$\left(1 - p_{j,\text{No Diabetes}}\right) = \frac{\bar{n}_j}{\bar{N}_j}$$
$$\left(1 - p_{j,\text{Diabetes}}\right) = \frac{n_j}{N_j} \, . \tag{8.7}$$

The first expression we derive relates the population mortality probability to the sum of individuals dying in that year divided by the sum of individuals alive at the beginning of the year:

$$\left(1 - p_{j,\text{Population}}\right) = \frac{n_j + \bar{n}_j}{N_j + \bar{N}_j} \, . \tag{8.8}$$

The second expression describes the relative risk of mortality for those with and without diabetes. This relative risk has been examined in multiple studies (CDC, 2011b; Emerging Risk Factors Collaboration, 2011). We use the estimate from the CDC (2011b) that individuals with diabetes have approximately twice the risk of mortality to obtain

$$\frac{n_a}{N_a} = 2 \frac{\bar{n}_a}{\bar{N}_a} \, . \tag{8.9}$$

Solving (8.8) and (8.9) for n_j and $\overline{n}_j$ and plugging into (8.7) give the mortality probabilities by diabetes state:

$$\left(1 - p_{j,\text{No Diabetes}}\right) = \frac{\left(\overline{N}_j + N_j\right)\left(1 - p_j\right)}{\overline{N}_j + 2N_j}$$

$$\left(1 - p_{j,\text{Diabetes}}\right) = \frac{2\left(\overline{N}_j + N_j\right)\left(1 - p_j\right)}{\overline{N}_j + 2N_j}. \tag{8.10}$$

With the multiattribute utility model, the decision analytic framework, and the initial data and calculations discussed, we now present an analysis of the consumption of SSB.

CASE EXAMPLE: VALUE TO THE INDIVIDUAL

A medical decision analysis can be used to derive numerous insights. In this section, we calculate the value of a decision to the decision-maker in terms of that decision's effect on his/her expected life.

The preferences are modeled using a multiattribute utility function as previously described. The change in risks is modeled using the decision tree of Figure 8.2 that illustrates the change in probabilities associated with each decision. The expected utility for each decision is calculated, and the change in consumption necessary for the expected utilities to be equal can be determined. This change in consumption to reach indifference determines the value in terms of lost life.

We present an illustrative example and conduct sensitivity analyses to derive insights. Consider Janet, a 25-year-old professional who estimates her constant annual consumption to be $80,000. Her risk tolerance is $40,000. She would be willing to forgo 1.5% of her income above bare survival in order for her life to be 1% longer.

With these parameters, we find that if Janet were to drink one sugar-sweetened soft drink per day for the rest of her life, she would value each drink at a maximum of $7.64. This value is found using the decision framework with $r = 1.83$ to correspond to the risks of daily sugar-sweetened soft drink consumption (Schulze et al., 2004). The difference in constant annual consumption necessary for the decision alternatives to have equal expected utility is $2789, which is averaged over each beverage to determine the value per drink. Because this relative risk is reported for one or more sugar-sweetened soft drink consumed per day, the value calculated is a maximum value per drink. If more than one is consumed per day, the value per drink decreases although the value per year remains constant. Table 8.2 reports the relative risks associated with different consumption rates and the value per drink for Janet.

We examine the sensitivity of these results to the elicited parameters. We first examine the effect of the trade-off parameter η when Janet consumes at least one SSB a day. The reported values are the maximum per beverage value. As the trade-off parameter η increases, the value per beverage increases as shown in Figure 8.4. This result is expected; the increase in value corresponds to the willingness to pay more for increases in life expectancy as indicated by the increase in η. Note this trend is not linear but is slightly convex.

Table 8.2   The per beverage value ranges calculated for the case example parameters

Sugar-sweetened beverage consumption	Relative risk of type 2 diabetes*	Values per beverage for case example
1–4/month	1.06	$4.74–$18.98
2–6/week	1.49	$5.55–$16.64
≥ 1/day	1.83	$0–$7.64

*As reported in Schulze et al. (2004).

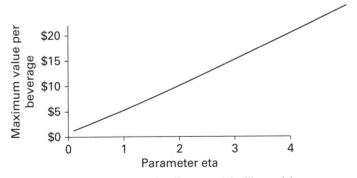

Figure 8.4   The effect of eta in the trade-off exponential utility model.

We next consider the effect of the relative risk of type 2 diabetes. The results are presented in Figure 8.5. From this figure, we can determine the value of other decisions to Janet that affect her risk of diabetes. For example, suppose Janet had a relative risk of type 2 diabetes of 1 and decided to begin a diet and exercise program that had been shown to reduce her risk of type 2 diabetes by 30%. The value of this program in terms of improving her mortality risk would be over $1200 per year.

We next consider risk tolerance and constant annual consumption. A decrease in risk tolerance results in an increase in the value per SSB, as shown in Figure 8.6. This trend makes intuitive sense; as an individual

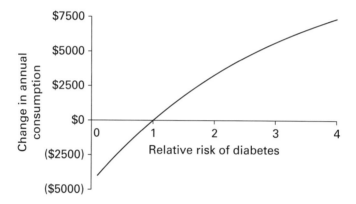

Figure 8.5 Effect of changes in relative risk of diabetes on consumption to compensate for risk.

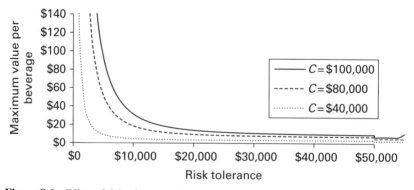

Figure 8.6 Effect of risk tolerance and consumption on the value of daily sugar-sweetened beverage consumption, $\eta = 1.5$.

becomes less willing to accept risk, any action that increases risk must have greater value.

A decrease in constant annual consumption leads to a decrease in the value per SSB. This result also makes intuitive sense because the preference model is based on percentage trade-offs between constant annual consumption and lifetime. For two individuals with the same percentage trade-off preference, the individual with a greater constant annual consumption is willing to forgo a greater amount of consumption for lifetime leading to placing a greater value, in absolute terms, on the same increase in risk.

SOCIETAL ANALYSIS

Thus far, our treatment of medical decisions has been on the level of an individual decision-maker. We now turn the discussion to conducting an analysis on the impact of a decision on society as a whole.

Medical decisions can impact society in a variety of ways. For example, the life years lived, the quality of health, or the medical costs incurred by society as a whole, as the result of a decision or set of decisions, can be examined. In this section, we illustrate how the decision tree framework already presented can be applied to such societal analysis. We continue the illustrative example on SSB in calculating the effect on expected medical costs. These costs may be paid by the individual, an insurance company, the government, or a combination of entities. Altogether, however, these medical costs represent a burden on society as a whole.

The outcomes represented in the decision tree must be described in terms that are relevant to society. For example, medical costs are a measure of the economic burden on society. In the case of diabetes, research has found that the average medical costs for an uncomplicated case of diabetes are $6649 per year (American Diabetes Association, 2008). The decision tree framework from Figure 8.2 is still used, but each outcome is described by medical costs instead of the multiattribute utility for preferences used in the case of the individual.

The calculations are conducted as follows. To calculate the expected medical costs due to diabetes, we must specify the year in which the individual is diagnosed with diabetes and the year in which the individual dies. Let the subscript i indicate the age at diagnosis of diabetes, and let the subscript j denote the age at death. The total average medical costs due to diabetes are the average annual cost multiplied by the years with diabetes:

$$\text{Total Cost}_{i,j} = (j - i) \text{ Annual Cost.} \tag{8.11}$$

The expected medical costs are determined by the summation of the product of (8.11) and the probability of diagnosis at age i and death at age j:

$$E[\text{Cost}] = \sum_j p(\text{Die}_j, \text{Diabetes}_i) \text{ Total Cost}_{i,j}. \tag{8.12}$$

In this formulation, we assume the annual costs are the average for an uncomplicated case of diabetes. This approach represents an underestimation of medical costs when complications due to diabetes occur.

We calculate the expected medical cost associated with a relative risk of type 2 diabetes of 1.83. For women, this is the relative risk associated with consuming at least one sugar-sweetened soft drink a day. We use the midpoint of each age range in the calculation. The results are presented in Figure 8.7. The lower expected cost among males is due to the lower risk of type 2 diabetes among males.

An individual's age plays a large role in the expected cost. Because we assume the decision-maker does not have diabetes at the time of the decision, increasing age means the individual has fewer life years remaining over which to accumulate potential medical costs.

To determine the effect of a decision on society at large, we require data on the decisions made by the population. The consumption of SSB by the Americans has been studied using the National Health and Nutrition Examination Survey that found that on average, 63% of the US adults consume an average of 17 ounces of SSB per day (Bleich et al., 2009). We can estimate the expected medical costs due to SSB by assuming this

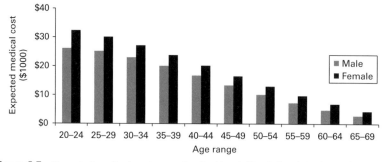

Figure 8.7 Expected medical costs associated with a 1.83 relative risk of type 2 diabetes.

rate is constant across ages and gender. We multiply the expected medical costs for each age group by the number of individuals in that age group. We use 5-year age groups, that is, ages 25–29, 30–33, and so on, as reported by the 2010 US census (Howden and Meyer, 2011). We use the expected medical costs corresponding to the midpoint of each group.

We find that the lifetime expected medical cost due to consuming SSB for adults ages 20 and over is $2.18 trillion. For the 80 years, it will take the youngest in this cohort to possibly reach age 100; this averages out to over $27 billion per year. This figure should be placed in the context of the US healthcare system as a whole. The US national health spending reached $2.7 trillion in 2011 (CMS, 2012), making the annual $27 billion per year represent approximately 1% of annual healthcare spending.

This analysis may also be placed in the context of other economic analyses of diabetes. In 2007, diabetes is estimated to have cost the United States $218 billion when both medical costs and lost productivity are included (Dall et al., 2010). The decision tree framework enables the quantification of the contribution of individual decisions to the burden on society. Analyses such as this highlight the significant national impact of individual decisions affecting wellness.

QUALITY OF HEALTH CONSIDERATIONS

We have discussed modeling preferences over length life and consumption and valuing decisions in terms of these preferences. In addition to length of life, however, people often also place value on quality of health or quality of life. This section discusses the quantification and modeling of quality of health.

Measuring Quality of Health

The concept of quality of health is straightforward to understand; some health states are preferred to others. But it has a less obvious measurement. Quantification can be accomplished through the comparison of perfect health to another health state, but multiple such approaches exist, including the standard gamble, the time-trade-off, and the rating scale method.

The standard gamble method relies on a comparison between a certain outcome and an uncertain deal. The decision-maker is asked to

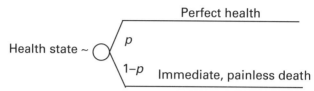

Figure 8.8 Using a standard gamble question to determine the utility of a health state.

consider the remainder of his/her life in the health state being measured. The decision-maker is then asked to consider an uncertain deal in which with probability p, he may obtain perfect health for the remainder of his/her life and with probability $(1-p)$, he will experience instant, painless death. Figure 8.8 illustrates the two prospects. The decision-maker is asked to determine the probability p that would make him just indifferent between the two. As the health state approaches perfect health, the probability p to obtain indifference approaches 1. As the health state becomes worse, the probability p will decrease.

The use of standard gambles follows from the definition of utility in decision analysis (von Neumann and Morgenstern, 1947). The nature of the uncertain deal means that preferences under uncertainty are captured. The element of uncertainty is important because it is inevitably present in medical decision-making situations. However, the presence of uncertainty can also contribute to difficulty for the decision-maker. The decision-maker may have difficulty assigning a probability p to health states close to perfect health as the probability is so close to 1. Other decision-makers may be unwilling to consider any risk of death for any health state, even if they are exposed to mortality risks on a daily basis. For these individuals, the time-trade-off method may be more appropriate.

The time-trade-off method assesses preferences for health states under conditions of certainty (Torrance, Sackett, and Thomas, 1972). The decision-maker is asked to consider the prospect of spending the rest of his/her life in a health state less than perfect health. Instead of this prospect, the decision-maker could spend the rest of his/her life in perfect health, but the length of life will be shorter. The decision-maker is asked to specify what reduction in life length would make his/her just indifferent between the two deterministic prospects. The time-trade-off utility is the ratio of time that would be spent in perfect health (life expectancy less reduction for indifference) divided by life expectancy.

A still simpler method is the rating scale. In this approach, death and perfect health lie at opposite ends of the scale. The decision-maker is asked to place the health state on the scale proportional to his/her relative preference for that state. This method does not incorporate uncertainty or trade-offs and is not supported by decision analytic theory. Due to these limitations as well as cognitive biases that can affect the results, this method is generally recommended for ordinal rankings or as a warm-up exercise only (Torrance, Feeny, and Furlong, 2001).

Preference Models with Quality of Health

Once a quantitative measure is placed on quality of health, a utility function or other models to describe preferences over quality of health may be specified. Multiple approaches exist to modeling preferences over quality of life, length of life, and monetary wealth. Some models only consider quality and length of life, while others also include monetary wealth. We describe how to use the trade-off utility function presented earlier in the chapter to include quality of health considerations and also review some of the other most well-known approaches.

Let q denote the quality of life measure. For the multiattribute utility function based on trade-offs between life and consumption, a quality of life measure can be included:

$$U(c,l,q) = 1 - \exp\left[-\frac{c}{\rho}\left(\frac{ql}{l}\right)^{\eta} \right]. \tag{8.13}$$

Because (8.13) is used to describe the remainder of the decision-maker's life and the quality of health may not be constant, a weighted sum of the years spent in each health state may be used. Such considerations could be used in a more specific analysis of SSB that included quality of health states associated with diabetes.

Another commonly used model for preferences over health states is the quality-adjusted life year (QALY) model. This model is often used in medical cost-effectiveness studies (Gold et al., 1996). In the QALY model, if the quality measure is constant for the remaining lifetime t, then the QALYs are calculated by the product

$$QALYs = tq. \tag{8.14}$$

If the quality measure changes over time periods, then each quality measure is weighted by the time spent in that health state and summed to give the QALYs. Note that in order for the QALYs to represent an individual's utility over health quality and life years, then q must accurately represent the decision-maker's utility for health quality and the following conditions must also hold (Pliskin, Shepard, and Weinstein, 1980):

1. Utility independence between life years and health quality
2. Constant proportional trade-off between health quality and life years
3. Risk neutrality over life years

In cases where these conditions do not hold, the QALY is a measure of time spent in health states of varying preference, but is not a utility.

Although the QALY model is frequently used, it is limited in its ability to describe preferences. For example, a person may wish to live long enough to see a child graduate from college. In this case, a life goal strongly influences this person's preferences over life and health. A modification to the QALY model has been proposed to model such situations (Hazen, 2007). A third attribute to represent goal attainment is introduced and denoted g. Let $u_Q(q)$, $u_T(t)$, and $u_G(g)$ denote the decision-maker's utility functions for the attributes health quality (q), life years (t), and goal attainment (g). Then a possible representation of utility to include an external goal is

$$U(q,t,g) = u_Q(q)u_T(t) + k_G u_G(g), \qquad (8.15)$$

where k_G is a weight representing the trade-off preference for goal attainment. Work has also been done to incorporate external goals in the elicitation of health quality measures (van der Pol and Shiell, 2007).

Another modification to the QALY model includes a term for monetary wealth or consumption in each time period (Smith and Keeney, 2005). We will denote the measure from this model as HC-QALY because it is a measure of both health and consumption. In this formulation, a utility function for the monetary wealth in each time period is multiplied by the quality of health for that period:

$$\text{HC-QALYs} = \sum_{t=1}^{N} q_t u_t(c_t), \qquad (8.16)$$

where the subscript t denotes each of N total time periods of equal length, q_t is the quality of health measure, c_t is the consumption, and $u_t(c_t)$ is the utility function over consumption in time period t.

CONCLUSION

Medical decision-making covers a variety of decision situations that affect health and the provision of medical care. This chapter has presented medical decision-making from a decision analytic perspective and has focused on modeling preferences with a multiattribute utility function and using this model within a decision tree framework to conduct analyses.

An illustrative example on the consumption of SSB has been used throughout to illustrate the concepts. While we use this specific example, the concepts and approach in the analysis can be applied to numerous decision situations.

The impact of medical decisions on both individuals and society can be analyzed. The outcomes within the decision tree must be described in terms of the appropriate attributes. For the individual, these attributes are those in the multiattribute utility function that determine his/her preferences. For society, these attributes are the ones relevant to society for which a quantitative measure is possible.

Through the proper application of decision analytic techniques, we can conduct analyses to better understand medical decisions and their impact on both individuals and society as a whole.

REFERENCES

Abbas, A.E., R.A. Howard. 2005. Attribute dominance utility. *Decision Analysis*. **2**(4): 185–206.

American Diabetes Association. 2008. Economic costs of diabetes in the U.S. in 2007. *Diabetes Care*. **31** 596–615.

Arias, E. 2010. Unites States Life Tables, 2006. *National Vital Statistics Reports*. **58** 1–40.

Beauchamp, T. L., J. F. Childress. 2001. *Principles of Biomedical Ethics*. 5th ed. Oxford University Press, New York.

Bleich, S. N., Y. C. Wang, Y. Wang, S. L. Gortmaker. 2009. Increasing consumption of sugar-sweetened beverages among US adults: 1988–1994 to 1999–2004. *American Journal of Clinical Nutrition*. **89** 372–381.

Centers for Disease Control and Prevention (CDC). 2011a. *Diabetes Data & Trends*. http://www.cdc.gov/diabetes/statistics/prevalence_national.htm (accessed August 29, 2014).

Centers for Disease Control and Prevention (CDC). 2011b. *National Diabetes Fact Sheet: National Estimates and General Information on Diabetes and Prediabetes in the United States*. U.S. DHHS, CDC, Atlanta, GA.

Centers for Medicare and Medicaid Services (CMS). 2012. *National Health Expenditure Tables.* http://www.cms.hhs.gov/NationalHealthExpendData (accessed August 29, 2014).

Dall, T. M., Y. Zhang, Y. J. Chen, W. W. Quick, W. G. Yang, J. Fogli. 2010. The economic burden of diabetes. *Health Affairs.* **29** 297–303.

Gold, M. R., D. L. Patrick., G. W. Torrence, D. G. Fryback, D. C. Hadorn, M. S. Kamlet, N. Daniels, M. C. Weinstein. 1996. Identifying and valuing outcomes. M. R. Gold, J. E. Siegel, L. B. Russell, M. C. Weinstein, eds. In *Cost-Effectiveness in Health and Medicine* (pp. 82–134). Oxford University Press, New York.

Hazen G. B. 2007. Adding extrinsic goals to the quality-adjusted life year model. *Decision Analysis.* **4** 3–16.

Howard, R. A. 1984. On fates comparable to death. *Management Science.* **30** 407–22.

Howard, R. A., A. E. Abbas. 2015. *The Foundations of Decision Analysis.* Forthcoming. Pearson Publishers, New York.

Howden, L. M., J. A. Meyer. 2011. Age and sex composition: 2010. *2010 Census Briefs.* http://www.census.gov/prod/cen2010/briefs/c2010br-03.pdf (accessed August 29, 2014).

Keeney, R. L. 2008. Personal decisions are the leading cause of death. *Operations Research.* **56** 1335–1347.

Narayan, K. M., J. P. Boyle, T. J. Thompson, S. W. Sorensen, D. F. Williamson. 2003. Lifetime risk for diabetes mellitus in the United States. *Journal of the American Medical Association.* **290**(14): 1884–1890.

von Neumann, J., O. Morgenstern. 1947. *Theory of Games and Economic Behavior.* 2nd ed. Princeton University Press, Princeton, NJ.

Pliskin, J. S., D. S. Shepard, M. C. Weinstein. 1980. Utility functions for life years and health status. *Operations Research.* **28** 206–224.

van der Pol, M., A. Shiell. 2007. Extrinsic goals and time tradeoff. *Medical Decision Making.* **27** 406–413.

Savulescu, J. 2007. Autonomy, the good life, and controversial choices. R. Rhodes, L. Francis, A. Silvers, eds. In *The Blackwell Guide to Medical Ethics* (pp. 17–37). Blackwell Publishing, Malden, MA.

Schulze, M. B., J. E. Manson, D. S. Ludwig, G. A. Colditz, M. J. Stampfer, W. C. Willett, F. B. Hu 2004. Sugar-sweetened beverages, weight gain, and incidence of type 2 diabetes in young and middle-aged women. *JAMA.* **292** 927–934.

Smith J. E., R. L. Keeney. 2005. Your money or your life: A prescriptive model for health, safety, and consumption decisions. *Management Science.* **51** 1309–1325.

Torrance, G. W., D. Feeny, W. Furlong. 2001. Visual analog scales: Do they have a role in the measurement of preferences for health states? *Medical Decision Making.* **21** 329–334.

Torrance, G. W., D. Sackett, W. Thomas. 1972. A utility maximization model for evaluation of health care programmes. *Health Services Research.* **7**(2) 118–133.

The Emerging Risk Factors Collaboration. 2011. Diabetes mellitus, fasting glucose, and risk of cause-specific death. *The New England Journal of Medicine.* **364** 829–841.

CHAPTER 9

Electric Power Vulnerability Models: From Protection to Resilience

Sinan Tas[1] and Vicki M. Bier[2]
[1] *Information Sciences and Technology, Penn State-Berks, Reading, PA, USA*
[2] *Department of Industrial and Systems Engineering, University of Wisconsin-Madison, Madison, WI, USA*

Prior to the terrorist attacks of September 11, 2001, the term "vulnerability assessment" was often used in relation to risk and system safety (Einarsson and Rausand, 1998). Concern about intentional threats was limited largely to the field of information security (e.g., Denning, 1999). Following September 11, however, there has been an increased emphasis on vulnerability to security threats from intelligent adversaries in all critical infrastructure systems, including electric power networks.

The United States created the Office of Homeland Security and the Homeland Security Council in 2001, which later led to the creation of the US Department of Homeland Security (DHS) in 2002. The first strategic document on homeland security, *The National Strategy for Homeland Security* (Bush, 2002), defined three strategic objectives: to prevent terrorist attacks against homeland targets, to reduce vulnerability to terrorism, and to minimize damage and

Breakthroughs in Decision Science and Risk Analysis, First Edition.
Edited by Louis Anthony Cox, Jr.

recover from attacks. These three objectives roughly correspond to threat, vulnerability, and consequence.

In 2003, the DHS published *The National Strategy for the Physical Protection of Critical Infrastructures and Key Assets* (DHS, 2003), a guide to the protection of critical infrastructures (e.g., water, energy) and key assets (e.g., national icons, nuclear power plants). This document highlights the importance of protection, response, and recovery and specifies the roles of the federal government and the states in identifying and securing the critical infrastructures and key assets under their control.

More recently, the *National Infrastructure Protection Plan* was written to provide additional guidance on how to make the nation's infrastructure safer, more secure, and more resilient (DHS, 2009). In particular, the plan requires implementation of a "long-term risk management program" that includes hardening, distributing, diversifying, and increasing the resilience of infrastructure against threats and hazards; interdicting potential attacks; and planning for rapid response to disruptions and rapid recovery.

The DHS defines vulnerability as a "physical feature or operational attribute that renders an entity, asset, system, network, or geographic area open to exploitation or susceptible to a given hazard" (DHS, 2010). In this definition, vulnerability could include any weakness that a terrorist could exploit or that makes the system more susceptible to either natural or man-made hazards. Other organizations use a narrower definition of vulnerability. For example, the US Coast Guard (USCG, 2003) defines vulnerability as "the conditional probability of success given a threat scenario occurs."

In this work, we adopt a broader definition of vulnerability, which also includes weaknesses in the short-term response of the system to outages or attacks (in the form of cascading failures) and also the long-term recovery of the system (i.e., restoration time). We define resilience as "the performance of a system over time in response to adverse change." Note that this definition is also consistent with the recent recommendation of the National Research Council (NRC) that DHS's vulnerability analyses should ideally address issues of system capacity and long-term adaptation (NRC, 2010).

In particular, with cascading failures, even small outages or attacks can have a large impact. Cascading failures have historically been considered a major unsolved problem for complex networks such as electricity systems, but recent developments in probabilistic analysis of cascading failure are making it possible to take cascading failures into account in methods of vulnerability assessment. Moreover, methods of vulnerability analysis can also be designed to highlight those vulnerabilities that are likely to lead to disproportionately long restoration times.

Protecting a significant fraction of a large, complex electric power network is typically not cost-effective and may even be infeasible (Bier et al., 2007; Tas, 2012). As an alternative, vulnerability models can aim to make electric power networks more resilient. Therefore, after discussing vulnerability-assessment methods that focus mainly on protection, we move on to discuss models of cascading failure and restoration times, with the goal of addressing the entire impact of a technological failure, natural disaster, or intentional attack (rather than just the immediate impact of such adverse events).

VULNERABILITY-ANALYSIS METHODS

Methods for vulnerability analysis include rating-based methods, risk-based methods, and game-theoretic methods. We discuss all three of these approaches in the following sections.

Note also that most risk-based and game-theoretic methods attempt to represent the physical system being analyzed in one of two different ways, using either topological models or flow-based models. Flow-based models aim to represent how a system actually functions. By contrast, topological models consider only the network structure. Thus, topological models can identify redundancies, potential bottlenecks, and so on, but cannot take into account factors such as capacity constraints (e.g., whether a particular line has sufficient capacity to serve all needed loads when other parts of the system have been degraded). We discuss rating-based methods, risk-based methods, and game-theoretic methods in turn in the following text.

Rating-Based Methods

Rating-based methods assign scores to various attributes of the system being analyzed. They are not specific to electric power networks and can generally be applied to a wide variety of systems, facilities, or networks. One such method is the Criticality, Accessibility, Recuperability, Vulnerability, Effect, and Recognizability (CARVER) method, originally developed by the US Special Operations Forces to help prioritize targets during the Vietnam War. The DHS uses CARVER to prioritize critical components and assets as part of the Buffer Zone Protection Plan (Bennett, 2007). In this method, each potential target is scored on the attributes of criticality, accessibility, recuperability, vulnerability, effect, and recognizability; the resulting numbers are simply added to find a final score for each target, effectively assigning the same weight to each attribute. By contrast, Ezell (2007) uses an additive preference model in which different weights are assigned to the various vulnerability attributes to quantify the importance of vulnerabilities.

A similar rating-based method is the Enhanced Critical Infrastructure Protection (ECIP) program, developed by Argonne National Laboratory for the DHS (Fisher and Norman, 2010). In this model, facilities fill out questionnaires in order to score their vulnerabilities in areas such as physical site security, security management, and so on. The weights corresponding to the various areas were assessed using expert elicitation and are hardwired into the program. The program is intended for use by facility owners and operators to prioritize protective measures for a facility and compare risk management controls (fences, cameras, etc.) with those of other facilities in the same sector.

Rating-based methods are easy to implement, are applicable to virtually any type of infrastructure system, and can provide useful insights to decision-makers. However, they are perhaps best utilized for qualitative tasks, such as identifying threat scenarios or screening critical components, since they generally do not include a realistic representation of the physical system being analyzed and hence cannot account for factors such as flows within the system, dependencies among components, and so on. This weakness is especially limiting when applying rating-based methods to complex networks such as electric power systems (as opposed to simple facilities that can be modeled as single entities), because the network topologies and dynamic behaviors of complex systems cannot be completely captured by rating the individual components of the system. Moreover, in most rating-based models, threat is not considered explicitly.

Risk-Based Methods

Probabilistic risk assessment has been used to analyze the vulnerabilities of infrastructure systems since the mid-1970s (Rasmussen, 1975). Many risk-based vulnerability-assessment methods were originally developed for assessing system safety and reliability and are well accepted for that purpose; by contrast, the application of risk-based methods to intentional security threats is newer and more controversial.

Risk-based models attempt to answer three fundamental questions (Kaplan, 1997):

1. What can go wrong?
2. How likely is it?
3. What are the consequences?

The resulting estimate of risk is generally expressed in the form

$$Risk = f(threat, vulnerability, consequence)$$

Most commonly, risk-based models compute risk as the product of threat, vulnerability, and consequence. Some of the most prominent risk-based vulnerability models are discussed in the following text.

There are numerous government-sponsored vulnerability-assessment methods based on risk. For example, Risk Analysis and Management for Critical Assets Protection (RAMCAP, 2006), developed by the American Society of Mechanical Engineers under sponsorship by the DHS, models vulnerabilities using event trees. Moreover, threat is estimated as a function of both target attractiveness and adversary capability and intent. Risk is then estimated as the product of threat, vulnerability, and consequence.

Similarly, the Transit Risk Assessment Module (TRAM) also uses event trees to assess vulnerability (U.S. Federal Emergency Management Agency, 2011). While TRAM was originally designed to prioritize surface-transportation assets for protection from possible threats, similar models can also be applied to other types of infrastructure. One such example is the Maritime Security Risk Analysis Model (MSRAM), developed by the US Coast Guard in 2006 to help prioritize the risks of terrorist attacks on ports and waterways (Parfomak and Frittelli, 2007).

Some government models are specific to a particular type of threat, such as DHS's Bioterrorism Risk Assessment model (DHS, 2006). This model uses event trees to prioritize bioterrorism threats based on subjective estimates of their probabilities and consequences. Similarly, the DHS also developed the Chemical Terrorism Risk Assessment model and an Integrated Chemical, Biological, Radiological, and Nuclear Assessment model (NRC, 2010).

Other government vulnerability-assessment tools and programs are designed for specific sectors. Examples include the Aviation Domain Risk Assessment, the Dams Sector Analysis Tool, the Emergency Services Self-Assessment Tool, the National Transportation Sector Risk Assessment, the Ports and Waterways Safety Assessment, the Risk Assessment Methodology for Water Utilities, and the Water Infrastructure Simulation Environment (DHS, 2010).

Risk-based models have also been frequently discussed in the academic literature. For example, Ezell, Farr, and Wiese (2000) propose a risk-based model to identify the vulnerable components of an infrastructure system. The model first identifies vulnerabilities and threats. For each of the most plausible threat scenarios, components are ranked according to how vulnerable they are to that threat (based on exposure to the threat and accessibility of the component by the attacker). Event trees are then used to identify and quantify the likelihood of sequences that could lead to adverse consequences.

Apostolakis and Lemon (2005) develop a risk-based approach to analyze the vulnerabilities of water, natural gas, and electric power distribution networks in the face of relatively minor terrorist attacks. The authors focus on the topological structures of the networks and also the geographic locations of the components. For example, they note that electrical service ducts are often collocated with (or geographically in proximity to) natural gas and water networks, creating critical points that are highly vulnerable. The model identifies all possible combinations of failures that may result from a single attack somewhere in the system, but without any assessment of their likelihood. Instead, using expert judgment, the authors estimate the accessibility of each critical point. The screening methodology then identifies and ranks failure combinations (i.e., minimal cut sets) based on their susceptibility to attack and the value of each target to the decision-maker (as calculated using multiattribute utility theory). For extensions of this work, see Koonce, Apostolakis, and Cook (2008) and Patterson and Apostolakis (2008).

Donde et al. (2005) use a graph-partitioning algorithm to identify critical power lines whose failure may cause severe system disruptions. Similarly, Lesieutre et al. (2006) use graph theory to identify subgraphs that are at risk for unmet demand in the case of extreme events. Bienstock and Mattia (2007) develop a graph-theoretic network model to explore how the robustness of a network can be improved at minimum cost. They consider two types of investments: adding more capacity to an arc and adding more arcs in parallel. He et al. (2004) analyze voltage stability to identify weak components, while Vulkanovski, Cepin, and Mavko (2009) generate fault trees for each load in a system and identify the most important elements in those fault trees using risk importance measures.

In their flow-based model, Bienstock and Verma (2009) use mixed-integer and nonlinear models to determine whether there is small number of arcs whose removal will cause a blackout. Similarly, Pinar et al. (2010) use a bilevel integer program to identify small groups of lines in a network whose removal would be anticipated to cause a severe disruption. They avoid the nonlinearity in their original bilevel mixed-integer nonlinear problem by approximating the problem as a mixed-integer linear problem. In the outer loop of their optimization, they identify the critical lines, while in the inner loop they measure blackout severity by solving the load-shedding problem that minimally decreases load given an assumed loss of the identified critical lines. Once the critical components that could lead to a blackout have been identified, the authors then find the minimum increase in generation required to avoid the blackout.

In general, many risk-based models are simple and practical to use; Ezell, Farr, and Wiese (2010) argue that the required inputs to risk-based models can be readily obtained through expert elicitation and note that PRA has been successfully applied to a number of large complex systems. However, Cox (2008) describes some of the limitations of risk-based models when applied to intentional threats. In particular, he notes that threat probabilities may not be well-defined constants, since an adversary might respond to any observed defenses; he also raises similar concerns about the possible ambiguity of vulnerability and consequence. Brown and Cox (2011) similarly warn that risk-based methods may result in misleading recommendations regarding protective actions, since the attack probabilities assumed in such methods may not reflect the attacker's ability to learn from the defender's analysis and/or the observed defenses.

The NRC (2010), in a recent review of DHS's approach to risk analysis, has stated that the basic idea of representing risk as a function of threat, vulnerability, and consequence is sound but recommends improvements to the validity and reliability of such models. For that reason, the NRC recommended that the DHS incorporate game theory into its vulnerability-analysis methods.

Game-Theoretic and Quasi-Game-Theoretic Methods

Unlike risk-based methods, game-theoretic vulnerability-assessment methods focus on the behavior of a strategic adversary. In particular, many game-theoretic vulnerability-assessment methods for networks are based on interdiction models. These games aim to determine how the attacker can "interdict" various components of the network in order to best achieve an objective. Then, the defender determines how best to operate the remaining network. In optimization terminology, these types of games (which are special cases of sequential or Stackelberg games) are often represented as mixed-integer bilevel programs; see Wood (1993) for an overview of interdiction models. Most interdiction models in the literature are deterministic. However, see Cormican, Morton, and Wood (1998) for an overview of stochastic interdiction models. More recently, Janjarassuk and Linderoth (2008) reformulate stochastic network-interdiction problems as deterministic mixed-integer programs, and Morton, Pan, and Saeger (2007) apply stochastic interdiction to the problem of nuclear smuggling.

Interdiction models have been extensively applied to transportation, nuclear smuggling, border patrol, and so on. Because this field is so broad, we limit the remainder of our discussion in this section to models that are either applied or potentially applicable to electric power networks.

In addition to truly game-theoretic models, however, we also address models that use the so-called conversational games (i.e., "advice, suggestions, and counsel about how to think strategically"; Shubik, 1987) or worst-case assumptions regarding threat scenarios. These quasi-game-theoretic models typically do not include any consideration of optimal defenses and may not even have been intended as models of adversary behavior. However, they go beyond simple risk-based models, since their use of worst-case assumptions in selecting which threat scenarios to consider can still help to shed light on possible attacker behavior.

We begin by studying topological models. Albert, Albert, and Nakarado (2004) develop a topological model to study the structural vulnerability of the North American power grid. They compare the impact of various interdiction strategies, such as removing transmission substations at random, removing those nodes with the highest number of arcs into and out of them (i.e., the nodes of highest "degree"), or removing nodes in decreasing order of estimated load (where load is estimated by the number of paths through each node: i.e., "node betweenness"). They find that even the removal of a single transmission node can cause significant connectivity losses and that load-based or degree-based removal typically has much greater impact than removal of nodes at random.

In their topological model, Al-Mannai and Lewis (2008) use a game-theoretic approach, in which the defender minimizes the total network risk, calculated as the sum of the risks (vulnerability multiplied by consequence) of all components, where vulnerability is computed as a function of the attacker and defender resource allocations. Similar to Albert, Albert, and Nakarado (2004), Lewis (2009) attempts to correlate the vulnerability of a network with its topology by considering the degree of each node in the network, speculating that networks will generally be more vulnerable to the removal of higher-degree nodes. Based on the model of Al-Mannai and Lewis (2008), Lewis recommends that the defender allocate its resources to the most critical components but notes that the attacker's optimal strategy may therefore be to attack less critical but undefended components.

Holmgren (2007) use a topological model to analyze effective strategies for defending electric power networks against intelligent attackers. In their model, the defender can either harden components or decrease their recovery times. They conclude that the optimal trade-off between these two measures depends on both the defender's total level of resources and the nature of the attack scenario. For example, in the case of severe attack scenarios that are likely to cause large consequences, they find that for their assumed parameter values, most of the available defensive resources should be allocated to recovery rather than hardening.

However, many scholars (especially power system engineers) have pointed out the drawbacks of using topological models to analyze network vulnerabilities. In particular, Hines, Cotilla-Sanchez, and Blumsack (2010) find that power grids are generally more vulnerable to flow-based attacks that consider actual flows within the network than to attacks that consider only the topology of the network (such as the degree of each node). Therefore, we now consider flow-based models.

Salmeron, Wood, and Baldick (2004) develop a flow-based interdiction model to protect against worst-case attacks on electric transmission systems. The model is solved as a sequential game, in which the attacker selects an interdiction plan to maximize the cost of operating the network (including the cost of any lost loads), while the defender then operates the remaining parts of the network so as to minimize that cost. The authors solve the resulting optimization problem by a decomposition-based heuristic algorithm. Salmeron, Wood, and Baldick (2009) improve on that algorithm, with the result that they can generate faster and better solutions for considerably larger electric power grids.

Arroyo and Galiani (2005) reformulate the model in Salmeron, Wood, and Baldick (2004) as a general nonlinear mixed-integer bilevel programming problem, making it possible for the attacker and the defender to have different objective functions. Instead of the decomposition-based heuristic used by Salmeron, Wood, and Baldick (2004, 2009), Motto, Arroyo, and Galiana (2005) transform Salmeron's mixed-integer bilevel program into a mixed-integer nonlinear program using the duality theorem and then convert this new problem into a mixed-integer linear program. Using a flow-based model, Yao et al. (2007) extend Salmeron's problem to a trilevel sequential game (i.e., a defender–attacker–defender game) in which the defender is able to anticipate the optimal attack strategy for any given network structure and design the network accordingly. They also propose a solution procedure for the resulting game.

Bier et al. (2007) use a simple flow-based heuristic interdiction model, in which a greedy attacker interdicts the components with the maximum flow. In this model, there are three nested algorithms. First, the power flow in the network is simulated using a DC load-flow algorithm (Carreras et al., 2002) that minimizes the cost of operating the system. A greedy interdiction algorithm identifies the most heavily loaded line and sets its flow to zero (representing a hypothetical attack); the resulting flows in the rest of the network are then computed. Finally, a hardening algorithm identifies a set of potentially interdicted lines to be protected, as a way of assessing the effectiveness of protection against a greedy attacker. Bier et al. (2007) obtain results similar to those of Salmeron, Wood, and Baldick (2004), and note that hardening even a significant fraction of the transmission lines in a network may not be sufficient to substantially diminish the unmet demand resulting from a greedy attack, concluding that hardening of components is unlikely to be cost-effective. Tas (2012) extends the model of Bier et al. (2007) to include nodes as well as arcs, making it possible to model attacks against generators, loads, and transformers in addition to transmission lines. This approach allows consideration of various types of attacker and defender strategies, for example, strategies in which the attackers restrict their attention to a specific component type.

Finally, in their flow-based model, Romero et al. (2012) study the problem of allocating a fixed budget to minimize the consequences of an intelligent attack. In order to find an optimal defense strategy, they use Tabu search with an embedded greedy algorithm to simulate the attacker.

Game-theoretic models incorporate the intelligent nature of the terrorist threat, reflecting the fact that attackers can observe and investigate the potential vulnerabilities of a network. Moreover, flow-based game-theoretic models enable the attacker to consider the traffic on the network when planning an attack.

However, game-theoretic models also have some drawbacks. For example, it may be unrealistic to assume that attackers are perfectly rational and have unlimited computational ability. Another concern is the conservatism of game-theoretic models in assuming that the attacker will maximize the consequences of an attack (Ezell, Farr, and Wiese, 2010), which will result in defending against only the most severe attacks and may therefore leave the defender vulnerable to less severe attacks.

In addition, rating-based models and models based on system topology instead of flow may be too simplistic to provide realistic representations of network dynamics. On the other hand, some game-theoretic models that simulate power flows, such as the model of Salmeron, Wood, and Baldick (2004), are computationally demanding and may therefore be difficult to extend to represent some real-world concerns. For example, with the exception of Hines, Cotilla-Sanchez, and Blumsack (2010), Tas (2012), and Tas and Bier (2014). existing game-theoretic and quasi-game-theoretic models do not address the impact of cascading failures. In fact, other than Tas (2012), the only model we have identified that addresses both cascading failures and restoration times is that of Anghel, Werley, and Motter (2007), which does not include a game-theoretic representation of attacker behavior. In our view, one goal of vulnerability models should be to find an acceptable middle ground that is reasonably realistic in its representation of power flows but still simple enough to be readily applicable in practice and capable of handling cascading failure and restoration times without too much computational difficulty. A sufficiently comprehensive model also makes it possible to analyze and compare a wide range of protective measures to improve network resilience. Therefore, in the following sections, we review models of cascading failures and restoration times, respectively.

MODELING CASCADING FAILURES IN ELECTRIC POWER NETWORKS

Even small attacks or incidents can have a catastrophic impact on a system if there is a potential for cascading failure. Mili, Qiu, and Phadke (2004) explain cascading failure as follows: "the power that used to pass through the tripped lines finds its way through other links in the network, which in turn may overload some of them. …this sequence of line tripping followed by line overloading may propagate throughout the network until either the line overloading vanishes or the stability limits or voltage collapse limits are reached."

Pure topological models inherently cannot deal with cascading failure, because cascading failure depends on links or nodes being overloaded beyond their capacity, not just on the topology of the network. Thus, there are two ways to represent cascading failure in a network. One is to try to infer which components might experience high flows from their topological position in the network, while the alternative approach is to model the flows explicitly. As a result, as before, we classify the cascading models in the literature into two broad categories: topological models (generally within accurate hypotheses of how flow works) and more rigorous flow-based models.

Models of cascading failure can also be categorized as deterministic (where failure of an overloaded component is assumed to occur based on a deterministic condition, such as load exceeding capacity by a given percentage) or probabilistic (where failure of an overloaded component is assumed to occur at random or in an unpredictable manner). In either case, failure of overloaded nodes or arcs has the potential to result in cascading failures by causing other system components to become overloaded. We first review deterministic models of cascading failure and then consider probabilistic models of cascading failure.

Deterministic Models of Cascading Failure

In their topological model, Albert, Albert, and Nakarado (2004) simulate cascading failures deterministically by removing the ten nodes with the highest estimated loads, recalculating the estimated loads, and repeating the process until the estimated load shed is at least 60%. They note that removal of only 5% of the nodes in a system using this algorithm can result in predicted failure of almost the entire system. Moreover, they find much greater losses using this cascading-failure algorithm than using simpler load-based or degree-based algorithms that remove nodes based on their initial characteristics, without recalculation of loads.

Crucitti, Latora, and Marchiori (2004) develop a topological model that uses "the total number of most efficient paths" through a node as an indicator for the load served by that node. Cascading failure is assumed to occur when the load served by a given node is more than its pre-defined capacity, leading to recalculation of the loads at each remaining node, which could cause even more nodes to become overloaded. The authors find that under some circumstances, failure of even a single node can lead to a total blackout, especially if the original failed node had a high estimated load.

Zhao, Park, and Lai (2004) develop a topological model to analyze the vulnerability and tolerance of complex networks to cascading failures. In their model, the load carried by each node is approximated by the number of shortest paths passing through that node. The capacity of the node is in turn assumed to be proportional to the original load (e.g., 20% more than the original load). The node with the largest number of arcs (i.e., the highest degree node) is assumed to be attacked, leading to a new set of shortest paths. At that point, the node with the highest number of shortest paths is assumed to fail if it exceeds its capacity, with this process being repeated a predetermined number of times. Like Crucitti, Latora, and Marchiori (2004), Zhao et al. note that disabling one or a few nodes can result in a complete blackout through cascading failure, even if the nodes of the network have relatively high capacities.

Kinney et al. (2005) model the power grid as a weighted graph. In their model, cascading failures are represented dynamically. As in Zhao, Park, and Lai (2004), the number of paths through each nondisabled node (i.e., the node betweenness) increases as breakdowns occur (since other nodes are no longer usable), until the capacity of a node is exceeded, resulting in its failure. The authors assume that after a cascading failure, a previously overloaded component has the possibility of working again if the load later decreases to be below the capacity of the node. This study highlights the potential severity of small attacks targeted at nodes with either high node betweenness or high degree and finds that losing even a single transmission station may reduce the capacity of a network by up to 25%.

Wang and Rong (2009) develop a topological model to analyze the robustness of the US power grid. They develop a new method to estimate and redistribute load levels, rather than the commonly used node-betweenness measure. According to this method, each node is assigned a predetermined load level. If a node is attacked, its load is distributed to its neighboring nodes in a manner proportional to their loads. Wang and Rong try removing loads in both ascending and descending order of load. Surprisingly, they find that when the initial load on the system is small enough, attacks on the least heavily loaded nodes can actually be more harmful than attacks on heavily loaded nodes.

Dueñas-Osorio and Vemuru (2009) use a node-betweenness measure to estimate the load flows in a network and analyze the impact of the initial network design on the potential for cascading failure. They conclude that increasing the capacity of the network does not always increase its robustness to cascading failures and that other types of design changes (such as reducing congestion, making the network more decentralized, or increasing the number of alternative routes between any given origin and destination) can be more useful.

Buldyrev et al. (2010) develop a topological model to analyze the impact of cascading failure in two interdependent networks, such as an electric power network and a communication network that depends on it. The model randomly removes a fraction of nodes in one network and assumes cascading failure of the corresponding nodes in the other network (together with the edges that connect the failed nodes in the two networks), which can in turn cause new failures in the original network. The process can continue until there are either no more edges to remove or no more nodes to fail. The authors then examine how much of the supporting network must be protected so that the disabled nodes constitute only a small portion of both networks. The authors find that networks with a more variable degree distribution (i.e., with some high-degree nodes and some low-degree nodes) are generally less robust to random attacks, because failure of high-degree nodes can cause significant damage.

As noted earlier, topological models estimate the flows in a network based on the inherent structure of the network. However, an electric power system may experience different loads at different times depending on the system characteristics. Therefore, rigorous flow-based models have been developed to simulate how flows within the system change after some components have been disabled.

The Critical Infrastructure Protection Decision Support System (CIP/DSS) includes a submodel specifically for electricity systems. Jointly developed by Argonne National Laboratory, Los Alamos National Laboratory, and Sandia National Laboratories (Bush et al., 2005) for use by government and industry, CIP/DSS represents the functional dependencies within and among various infrastructure systems as flows and then simulates the dynamics resulting from these dependencies. However, it is extremely detailed and computation intensive and therefore may not be practical for routine industrial use.

Similarly, the Critical Infrastructure Modeling System (CIMS) was developed by Idaho National Laboratory in 2005 to identify interdependencies among various infrastructure sectors. CIMS uses discrete-event simulation to help visualize cascading failures and to explore the possible consequences of infrastructure interdependencies. Its main purpose is to conduct "what-if" analysis to understand the vulnerabilities of infrastructure systems (Dudenhoeffer et al., 2006).

Other models are specifically designed to simulate flows in power networks. Ni et al. (2003) develop an online flow-based model for use by operators to decide when to alleviate stresses on a transmission network. The model deterministically removes circuits if they pass the emergency-overload limit a specified number of times (e.g., once or twice) and then recalculates all flows.

The Transmission Reliability Evaluation of Large-Scale System (TRELSS) is a risk-based model that simulates the cascading process based on some predetermined initial events in order to identify and rank critical cascading scenarios based on their severity and likelihood. The methodology has been used in transmission system enhancement projects as a prioritization tool (Hardiman, Kumbale, and Makarov, 2004).

In their flow-based model, Zima and Andersson (2004) calculate the impact of line outages on the flows in the remaining lines and cascade any overloaded lines deterministically after each calculation. The authors also calculate the minimal changes and/or load shedding needed to mitigate cascading failure.

Hines, Cotilla-Sanchez, and Blumsack (2010) develop a flow-based model to calculate the impact of cascading failures on blackout size. In this model, after an initial failure, each neighboring component is assumed to be removed from service if its flow exceeds 50% of its capacity for more than 5 seconds; power flows are then recalculated. To our knowledge, this is the only flow-based model addressing adversarial threats that are explicitly designed to cause cascading failures, although other models could also be used for this purpose.

Despite the development of many deterministic models of cascading failures, cascading failures have historically been considered a major unsolved problem in complex networks such as electricity systems, since it has proven difficult to determine exactly where and when cascading failures will occur. In particular, deterministic flow-based models are incapable of considering the hidden unidentified failures that may lead to cascading failure, since by definition such latent failures are unobservable. Therefore, we now consider probabilistic models of cascading failure.

Probabilistic Models of Cascading Failure

Since attempts to replicate the physics of what goes on in a network have not been particularly successful, some authors have proposed using probabilistic approaches to account for the difficulty of predicting cascading failures. For example, in their topological model, Liao et al. (2004) compute the probability that a random outage will produce a cascading failure of a certain size, conditional on an assumed set of hidden failures and network stress levels.

Based on historical data, Mili, Qiu, and Phadke (2004) estimate the likelihood of cascading failure for a given type of relay based on the percentage of relays of that type involved in past cascading failures. They then use the resulting likelihoods to calculate the probability of system failure using event trees.

In their flow-based model, Chen, Thorp, and Dobson (2005) assume that an overloaded line is more likely to fail in its first exposure to overload than in subsequent exposures. This is consistent with the idea that cascading failures result from preexisting hidden faults.

Dobson et al. (2001), Carreras et al. (2004), Dobson et al. (2007), and Newman et al. (2011) propose a probabilistic flow-based model (called the OPA model) in which cascading failure occurs with some probability when one or more lines are at or near their maximum capacities. Their model has two intrinsic dynamics, slow and fast. The slow dynamics represent load growth and response to blackouts on a scale of days, months, or years. On each day of the simulation, the loads of the network are assumed to change by a factor of λ, where λ is a uniform random number between λ_{min} and λ_{max}, with mean value $\bar{\lambda}$ larger than one. Transmission-line capacity is also assumed to increase in response to blackouts.

The fast dynamics represent the possibility of cascading failures on a scale of seconds to minutes. The assumption is that even though disruptions can happen at any time, they are more likely to happen at or near times of peak load, when lines are highly stressed. Each overloaded line is assumed to fail with a specified probability p, after which loads are recomputed, with the process continuing until there are no more overloaded lines.

In order to model how transmission lines cascade and predict the total number of line failures, Dobson, Kim, and Wierzbicki (2010) propose using a probabilistic branching process, the parameters of which are estimated based on observed transmission-line failures in the past. Then, they test the closeness of the predicted distribution obtained using their branching process model with the distributions of the number of transmission-line failures obtained using their OPA model and find close results in most cases. Moreover, Dobson (2012) shows that line outages predicted by his branching process match well with 12.4 years of transmission-line outage data from a North American utility company.

Inspired by Dobson et al. (2001), Anghel, Werley, and Motter (2007) develop a probabilistic flow-based model that represents both cascading failure and the system operator's response to disruptions. The authors analyze the optimal trade-off between the risk of cascading failure and the losses due to intentional load shedding by the operator. Tas (2012) also uses a modified version of the model of Dobson et al. (2001) to represent cascading failure.

MODELING RESTORATION TIMES

Most of the models discussed earlier represent the estimated impact of a disruption in a static manner, as a snapshot of the system. However, system owners, operators, and customers also care about how long it will take for a system to return to normal operating conditions; likewise, intelligent adversaries may consider the likely durations of the disturbances they cause in deciding which components to target. Thus, models that consider the restoration times of failed components can give a more realistic portrayal of system risks. We again begin with topological models and then move on to flow-based models.

In their risk-based model, Apostolakis and Lemon (2005) use restoration times as part of their consequence analysis. However, they limit their analysis to minor attacks that would involve only minimal restoration times (e.g., less than a week), so their model may not be relevant to more serious threats. In particular, they note that coordinated attacks on several locations (which would require more time to repair) may also involve larger minimal cut sets and therefore may not be computationally feasible to analyze in their model.

Holmgren (2006) compares three strategies to decrease the vulnerability of power grid networks to both natural hazards and planned attacks: increasing the robustness of the network to cascading failures (by adding two underground power cables), increasing the ability of the network to recover quickly (a 15% reduction in restoration times), and increasing both robustness and rapid recovery (with one new underground cable and a 10% reduction in restoration times). He concludes that the combined strategy would yield roughly twice as much reduction in vulnerability as could be expected from either of the individual strategies. However, he also notes that a more realistic consequence analysis would require the use of a flow-based model.

Therefore, we now consider flow-based models involving restoration times. CIP/DSS simulates the impact of disruption over time using system dynamics (Bush et al., 2005) and can be used to analyze how quickly a system would recover based on various recovery scenarios (e.g., with two repair crews vs. three).

Salmeron, Wood, and Baldick (2004) weight the importance of each component in their model by the average time required for repair or replacement of that component type and use this information in anticipating which components would be most attractive to attackers. However, they do not explicitly simulate changes in system performance over time as a result of restoration efforts after an attack. Similarly, Tas (2012) weights the importance of each component by the average time required for its restoration. He further analyzes changes in attacker and defender behavior when restoration times are taken into account.

Romero et al. (2012) similarly use different restoration costs and times for each component type in their budget-allocation algorithms for system improvement; however, they do not address cascading failure.

Anghel, Werley, and Motter (2007) model the restoration time of a transmission line as having a constant minimum value, plus an exponentially distributed additional delay time. The authors simulate the resulting behavior of the system over time and analyze the optimal level of load shedding during the period before system restoration.

SUMMARY

Overall, game-theoretic methods of vulnerability analysis are preferred to rating-based or risk-based methods, when we aim to represent the behavior of a strategic threat. In particular, such models can represent threat as a function of the vulnerability of the network to an attack and the possible consequences of a successful attack.

In addition, while pure topological models may provide some insights into the structural changes necessary to achieve a less vulnerable system, modeling the flows within a network is critical to understanding what will happen in the event of a disruption. Flow-based models are therefore more realistic than topological models.

Cascading failures can be critical to understanding the response of complex systems that operate at or near their capacity. Hence, representing the possibility of cascading failure in electric power networks can help in identifying the most critical components. Exact modeling of the dynamics of cascading failure is not achievable at present; however, recent probabilistic approaches to modeling of cascading failure may provide a practical solution to this problem. Finally, we believe it is useful to incorporate restoration times into models of vulnerability, both to fully represent the overall impact of an attack and to capture the possible effects of restoration times on the attacker's choice of strategies.

Table 9.1 summarizes the literature on vulnerability analysis of electric power networks, based on the vulnerability methods they adopt, the models they use for cascading failure (if any), and whether they incorporate restoration times. As can be seen from the table, most models address only one or two aspects of the problem. Anghel, Werley, and Motter (2007) develop a risk-based flow model that represents cascading failure probabilistically and includes restoration times, but

Table 9.1 Comparison of power network vulnerability models in the literature

| No. | Literature | Vulnerability methods | | | | | Models of cascading failure | | | | Models of restoration times |
| | | Score-based models | Risk-based models | | Game-theoretic models | | Deterministic models | | Probabilistic models | | |
			Topological	Flow based	Topological	Flow based	Topological	Flow based	Topological	Flow based	
1	Anghel, Werley, and Motter (2007)			+						+	+
2	Albert, Albert, and Nakarado (2004)				+		+				
3	Apostolakis and Lemon (2005)		+								
4	Arroyo and Galiani (2005)					+					
5	Bienstock and Mattia (2007)		+								
6	Bier et al. (2007)					+					
7	Buldyrev et al. (2010)						+				
8	Carreras et al. (2004)								+		
9	CARVER (Bennett, 2007)	+									
10	CIMS (Dudenhoeffer et al., 2006)							+			+
11	CIP/DSS (Bush et al., 2005)							+			+
12	Crucitti, Latora, and Marchiori (2004)						+				

#	Reference								
13	Dobson et al. (2001, 2007)						+		
14	Donde et al. (2005)	+							
15	Dueñas-Osorio and Vemuru (2009)					+			
16	ECIP (Fisher and Norman, 2010)	+							
17	Ezell Farr, and Wiese (2000)	+							
18	Ezell (2007)	+							
19	He et al. (2004)	+							
20	Hines, Cotilla-Sanchez, and Blumsack (2010)		+	+	+	+			
21	Holmgren (2006)		+	+					+
22	Holmgren (2007)		+	+					+
23	Kinney et al. (2005)				+				
24	Lewis (2009)		+						
25	Lesieutre et al. (2006)	+				+			
26	Motto, Arroyo, and Galiana (2005)		+						
27	Newman et al., 2011)		+				+		
28	Pinar et al. (2010)	+							
29	RAMCAP (2006)	+							
30	Romero et al. (2012)	+					+		+

(continued)

Table 9.1 (Continued)

No.	Literature	Vulnerability methods					Models of cascading failure				Models of restoration times
		Score-based models	Risk-based models		Game-theoretic models		Deterministic models		Probabilistic models		
			Topological	Flow based	Topological	Flow based	Topological	Flow based	Topological	Flow based	
31	Salmeron, Wood, and Baldick (2004, 2009)					+					+
32	Tas (2012)					+				+	+
33	U.S. Federal Emergency Management Agency (2011)		+								
34	Vulkanovski, Cepin, and Mavko (2009)		+						+	+	
35	Yao et al. (2007)					+					

they do not consider the behavior of a strategic adversary. Salmeron, Wood, and Baldick (2004, 2009) and Holmgren (2007) use game-theoretic models with restoration times, but do not consider cascading failure. Hines, Cotilla-Sanchez, and Blumsack (2010) use a quasi-game-theoretic approach and model cascading failure; however, they do not consider restoration times. Dobson et al. (2001), Carreras et al. (2004), Dobson et al. (2007), and Newman et al. (2011) use flow-based models to model cascading failure probabilistically, but do not consider intentional threat scenarios or restoration times.

Tas (2012) uses a simple greedy heuristic attack and defense algorithm that allows modeling of both cascading failure and restoration times. He further analyzes how an attacker can choose attack strategies with the goal of causing cascading failures and/or long restoration times and discusses possible defenses against different types of attack strategies. Thus, this model fills an important gap in the literature, since it offers a simple heuristic approach suitable for widespread use by practitioners (unlike some of the more elaborate optimization-based models in the literature) while still reflecting important properties of real electricity networks (including realistic flows, cascading failures, and restoration processes). This approach is consistent with the view of Brown (2005) that an effective decision aid should "treat many issues minimally, rather than seek technical closure on any one." We believe that this approach (analyzing the entire impact of a disruption at a modest level of detail) will enable practitioners to measure network resilience more effectively than models that provide a greater level of detail but address only a few aspects of system performance.

REFERENCES

Albert, R., Albert, I., and Nakarado, G.L. (2004). "Structural vulnerability of the North American power grid." *Physical Review E*, 69(2), 025103.

Al Mannai, W.I. and Lewis, T.G. (2008). "A general defender-attacker risk model for networks." *The Journal of Risk Finance*, 9(3), 244–261.

Anghel, M., Werley, K.A., and Motter, A.E. (2007). "Stochastic model for power grid dynamics." In *40th Hawaii International Conference on System Sciences*, Big Island, HI, January 3–6.

Apostolakis, G.E. and Lemon, D.M. (2005). "A screening methodology for the identification and ranking of infrastructure vulnerabilities due to terrorism." *Risk Analysis*, 25(2), 361–376.

Arroyo, J.M. and Galiana, F.D. (2005). "On the solution of the bilevel programming formulation of the terrorist threat problem." *IEEE Transactions on Power Systems*, 20(2), 789–797.

Bennett, B.T. (2007). *Understanding, Assessing, and Responding to Terrorism: Protecting Critical Infrastructure and Personnel*, John Wiley & Sons, Inc., Hoboken, NJ.

Bienstock, D. and Mattia, S. (2007). "Using mixed-integer programming to solve power grid blackout problems." *Discrete Optimization*, 4(1), 115–141.

Bienstock, D. and Verma, A. (2009). "The N-k problem in power grids: New models, formulations and numerical experiment." *SIAM Journal on Optimization*, 20(5), 2352–2380.

Bier, V.M., Gratz, E.R., Haphuriwat, N.J., Magua, W., and Wierzbicki, K.R. (2007). "Methodology for identifying near-optimal interdiction strategies for a power network transmission system." *Reliability Engineering and System Safety*, 92(9), 315–323.

Brown, R.V. (2005). "Logic and motivation in risk research: A nuclear waste test case." *Risk Analysis*, 25(1), 125–140.

Brown, G.G. and Cox, L.A. Jr. (2011). "How probabilistic risk assessment can mislead terrorism risk analysts." *Risk Analysis*, 31(2), 196–204.

Buldyrev, S.V., Parshani, R., Paul, G., Stanley, H.E., and Havlin, S. (2010). "Catastrophic cascade of failures in interdependent networks." *Nature*, 464(7291), 1025–1028.

Bush, G.W. (2002). *The National Security Strategy of the United States of America*, Executive Office of the President, Washington, DC.

Bush, B., Dauelsberg, L.R., LeClaire, R.J., Deland D.R., and Samsa, S.M. (2005). "Critical infrastructure protection decision support system (CIP/DSS) project overview." Report no. LA-UR-05-1870, Los Alamos National Laboratory, Los Alamos, NM.

Carreras, B.A., Lynch, V.E., Dobson, I., and Newman, D.E. (2002). "Critical points and transitions in an electric power transmission model for cascading failure blackouts." *Chaos*, 12, 985–994.

Carreras, B.A., Lynch, V.E., Dobson, I., and Newman, D.E. (2004). "Complex dynamics of blackouts in power transmission systems." *Chaos*, 14, 643.

Chen, J., Thorp, J.S., and Dobson, I. (2005). "Cascading dynamics and mitigation assessment in power system disturbances via a hidden failure model." *International Journal of Electrical Power and Energy Systems*, 27(4), 318–326.

Cormican, K.J., Morton, D.P., and Wood, R.K. (1998). "Stochastic network interdiction." *Operations Research*, 46(2), 184–197.

Cox, L.A. Jr. (2008). "Some limitations of 'risk = threat × vulnerability × consequence' for risk analysis of terrorist attacks." *Risk Analysis*, 28(6), 1749–1761.

Crucitti, P., Latora, V., and Marchiori, M. (2004). "Model for cascading failures in complex networks." *Physical Review E*, 69(4), 045104.

Denning, D. (1999). *Information Warfare and Security,* ACM Press, Boston, MA.

DHS (2003). *The National Strategy for the Physical Protection of Critical Infrastructures and Key Assets,* U.S. Department of Homeland Security, Washington, DC.

DHS (2006). *Bioterrorism Risk Assessment,* U.S. Department of Homeland Security, Washington, DC.

DHS (2009). *National Infrastructure Protection Plan*, U.S. Department of Homeland Security, Washington, DC.

DHS (2010). *DHS Risk Lexicon*, Risk Steering Committee, U.S. Department of Homeland Security, Washington, DC.

Dobson, I. (2012). "Estimating the propagation and extend of cascading line outages from utility data with a branching process." *IEEE Transactions on Power Systems*, 27(4), 2146–2155.

Dobson, I., Carreras, B., Lynch, V., and Newman, D. (2001). "An initial model for complex dynamics in electric power system blackouts." In *Proceedings of the 34th Annual Hawaii International Conference on System Sciences* Maui, HI, January 3–6, pp. 710–718, IEEE Computer Society, Press, Washington, DC.

Dobson, I., Carreras, B.A., Lynch, V.E., and Newman, D.E. (2007). "Complex systems analysis of series of blackouts: Cascading failure, critical points, and self-organization." *Chaos*, 17, 026103.

Dobson, I., Kim, J., and Wierzbicki, K.R. (2010). "Testing branching process estimators of cascading failure with data from a simulation of transmission line outages." *Risk Analysis*, 40(3), 650–662.

Donde, V., Lopez, V., Lesieutre, B., Pinar, A., Yang, C., and Meza, J. (2005). "Identification of severe multiple contingencies in electric power networks." In *Proceedings of 37th North American Power Symposium*, Ames, IA, October 23–25.

Dudenhoeffer, D.D., Permann, M.R., and Manic, M. (2006). "CIMS: A framework for infrastructure interdependency modeling and analysis." In *Proceedings of 2006 Winter Simulation Conference*, pp. 478–485.

Dueñas-Osorio, L. and Vemuru, S.M. (2009). "Cascading failures in complex infrastructure systems." *Structural Safety*, 31(2), 157–167.

Einarsson, S. and Rausand, M. (1998). "An approach to vulnerability analysis of complex industrial systems." *Risk Analysis*, 18(5), 535–546.

Ezell, B.C. (2007). "Infrastructure Vulnerability Assessment Model (I-VAM)." *Risk Analysis*, 27(3), 571–583.

Ezell, B.C., Farr, J.V., and Wiese, I. (2000). "Infrastructure risk analysis model." *Journal of Infrastructure Systems*, 6, 114.

Fisher, R.E. and Norman, M. (2010). "Developing measurement indices to enhance protection and resilience of critical infrastructure and key resources." *Journal of Business Continuity and Emergency Planning*, 4(3), 191–206.

Hardiman, R.C., Kumbale, M., and Makarov, Y.V. (2004). "An advanced tool for analyzing multiple cascading failures", In *8th International Conference on Probabilistic Methods Applied to Power Systems*, Iowa State University, Ames, IA, September 12–16.

He, T., Kolluri, S., Mandal, S., Galvan, F., and Rastgoufard, P. (2004). "Identification of weak locations in bulk transmission systems using voltage stability margin index." In *International Conference on Probabilistic Methods Applied to Power Systems*, Ames, IA (pp. 878–882), September 12–16.

Hines, P., Cotilla-Sanchez, E., and Blumsack, S. (2010). "Do topological models provide good information about electricity infrastructure vulnerability?" *Chaos*, 20, 033122.

Holmgren, Å.J. (2006). "Using graph models to analyze the vulnerability of electric power networks." *Risk Analysis*, 26(4), 955–969.

Holmgren, Å.J. (2007). "A framework for vulnerability assessment of electric power systems." In *Reliability and Vulnerability in Critical Infrastructure: A Quantitative Geographic Perspective*, Murray, A. and Grubesic, T. (eds.). Springer, New York, 31–55.

Janjarassuk, U., and Linderoth, J. (2008). "Reformulation and sampling to solve a stochastic network interdiction problem." *Networks*, 52(3), 120–132.

Kaplan, S. (1997). "The words of risk analysis." *Risk Analysis*, 17(4), 407–417.

Kinney, R., Crucitti, P., Albert, R., and Latora, V. (2005). "Modeling cascading failures in the North American power grid." *The European Physical Journal B-Condensed Matter and Complex Systems*, 46(1), 101–107.

Koonce, A.M., Apostolakis, G.E., and Cook, B.K. (2008). "Bulk power risk analysis: Ranking infrastructure elements according to their risk significance." *International Journal of Electrica/Power and Energy Systems*, 30(3), 169–183.

Lesieutre, B.C., Roy, S., Donde, V., and Pinar, A. (2006). "Power system extreme event screening using graph partitioning." In *38th North American Power Symposium*, Southern Illinois University, Carbondale, IL, (pp. 503–510), September 17–19.

Lewis, T.G. (2009). *Network Science: Theory and Applications*, John Wiley & Sons, Inc., Hoboken, NJ.

Liao, H., Apt, J., and Talukdar, S. (2004). "Phase transitions in the probability of cascading failures." In Electricity Transmission in Deregulated Markets, conference at Carnegie Mellon University, Pittsburgh, PA, December 15–16.

Mili, L., Qiu, Q., and Phadke, A.G. (2004). "Risk assessment of catastrophic failures in electric power systems." *International Journal of Critical Infrastructures*, 1(1), 38–63.

Morton, D.P., Pan, F., and Saeger, K.J. (2007). "Models for nuclear smuggling interdiction." *IIE Transactions*, 39(1), 3–14.

Motto, A.L., Arroyo, J.M., and Galiana, F.D. (2005). "A mixed-integer LP procedure for the analysis of electric grid security under disruptive threat." *IEEE Transactions on Power Systems*, 20(3), 1357–1365.

Newman, D.E., Carreras, B.A., Lynch, V.E., and Dobson, I. (2011). "Exploring complex systems aspects of blackout risk and mitigation." *IEEE Transactions on Reliability*, 60(1), 134–143.

Ni, M., McCalley, J.D., Vittal, V., and Tayyib, T. (2003). "Online risk-based security assessment." *IEEE Transactions on Power Systems*, 18(1), 258–265.

NRC (2010). *Committee to Review the Department of Homeland Security Approach to Risk Analysis*, The National Academies Press, Washington, DC.

Parfomak, P.W. and Frittelli, J. (2007). *Maritime Security: Potential Terrorist Attacks and Protection Priorities*, Congressional Research Service, Washington, DC.

Patterson, S.A. and Apostolakis, G.E. (2008). "Identification of critical locations across multiple infrastructures for terrorist actions." *Reliability Engineering and System Safety*, 92(9), 1183–1203.

Pinar, A., Meza, J., Donde, V., and Lesieutre, B. (2010). "Optimization strategies for the vulnerability analysis of the electric power grid." *SIAM Journal on Optimization*, 20(4), 1786–1810.

RAMCAP (Risk Analysis and Management for Critical Asset Protection) Framework (2006). Available at http://onlinelibrary.wiley.com/doi/10.1002/9780470087923. hhs003/full.

Rasmussen, N.C. (1975). "Reactor safety study: An assessment of accident risks in US commercial nuclear power plants." WASH-1400, Nuclear Regulatory Commission, Washington, DC.

Romero, N., Xu, N., Nozick, L.K., and Dobson, I. (2012). "Investment planning for electric power systems under terrorist threat." *IEEE Transactions on Power Systems*, 27(10), 108–116.

Salmeron, J., Wood, K., and Baldick, R. (2004). "Analysis of electric grid security under terrorist threat." *IEEE Transactions on Power Systems*, 19(2), 905–912.

Salmeron, J., Wood, K., and Baldick, R., (2009). "Worst-case interdiction analysis of large-scale electric power grids." *IEEE Transactions on Power Systems*, 24(1), 96–104.

Shubik, M. (1987). "What is an application and when is theory a waste of time?" *Management Science*, 33(12), 1511–1522.

Tas, S. (2012). *A Comprehensive Method for Assessing the Resilience of Power Networks in the Face of an Intelligent Adversary* (doctoral thesis), University of Wisconsin-Madison.

Tas, S. and Bier, V. M. (2014). Addressing vulnerability to cascading failure against intelligent adversaries in power networks. Energy Systems, 1–21.

U.S. Coast Guard (2003). "Implementation of National Maritime Security Initiatives." *Federal Register*, 68(126), 39240–39250.

U.S. Federal Emergency Management Agency (2011). *Technical Assistance Catalog: Preparedness and Program Management Technical Assistance.* Available at http://www.fema.gov/pdf/about/divisions/npd/npd_technical_assistance_catalog.pdf (accessed October 10, 2014).

Vulkanovski, A., Cepin, M., and Mavko, B. (2009). "Application of the fault tree analysis for assessment of power system reliability." *Reliability Engineering & System Safety*, 94(6), 1116–1127.

Wang, J.W. and Rong, L.L. (2009). "Cascade-based attack vulnerability on the US power grid." *Safety Science*, 47(10), 1332–1336.

Wood, R.K. (1993). "Deterministic network interdiction." *Mathematical and Computer Modelling*, 17(2), 1–18.

Yao, Y., Edmunds, T., Papageorgiou, D., Alvarez, R. (2007). "Trilevel optimization in power network defense." *IEEE Transactions on Systems, Man and Cybernetics-Part C:Applications and Reviews*, 37(4), 712–718.

Zhao, L., Park, K., and Lai, Y.C. (2004). "Attack vulnerability of scale-free networks due to cascading breakdown." *Physical Review E*, 70(3), 035101.

Zima, M. and Andersson, G. (2004). "Wide area monitoring and control as a tool for mitigation of cascading failures." In *Eight International Conference on Probability Methods Applied to Power Systems*, Ames, IA, September 12–16.

CHAPTER 10

Outthinking the Terrorists

Jason Merrick[1], Philip Leclerc[1], Hristo Trenkov[2], and Robert Olsen[2]
[1] *Department of Statistical Sciences and Operations Research, Virginia Commonwealth University, Richmond, VA, USA*
[2] *Deloitte Consulting, Center for Risk Modeling and Simulation, Washington, DC, USA*

INTRODUCTION

Deciding how to defend against terrorist threats requires the best insights that decision and risk analysis methods can offer. Well-established decision analytic tools, such as decision trees, influence diagrams, and Bayesian networks, are commonly used in modeling counterterrorism decisions. These techniques use probabilities to model the uncertainty inherent in the domain, including uncertainty about terrorist attack methods and targets. This chapter reviews three approaches to determining which terrorist plans are most likely: directly eliciting subjective probabilities from intelligence experts, modeling adaptive decision-making and game-theoretic strategizing by intelligent foes, and text mining of intelligence data to see what is actually happening. We focus on two example counterterrorism decisions to better understand the problem: defending commercial aviation against surface-to-air missiles and screening cargo containers for nuclear threats.

von Winterfeldt and O'Sullivan (2006) perform a decision analysis to minimize the risk of attacks on commercial airlines using surface-to-air

Breakthroughs in Decision Science and Risk Analysis, First Edition.
Edited by Louis Anthony Cox, Jr.

missiles, called man-portable air defense system (MANPADS). In particular, they assessed the effectiveness of directed infrared counter-measures (DIRCMs). Figure 10.1 illustrates an influence diagram for the MANPADS countermeasure decision. Square nodes represent decisions, and round nodes represent uncertain events. The arrows indicate that the source node is relevant to the outcome of the destination node. Arrows pointing into a decision node indicate that the outcomes of their source nodes are known at the time the decision is made and are relevant to that decision. Arrows pointing into uncertainty nodes indicate that the probability distribution of the uncertainty is conditional on the outcome of the source node. The row of uncertainty nodes in the middle of Figure 10.1 is the observable chain of events in a MANPADS attack. These uncertainties depend on the bottom row of uncertainties that represent events that will affect the outcome of the attack whether countermeasures are in place or not and the top row of uncertainties

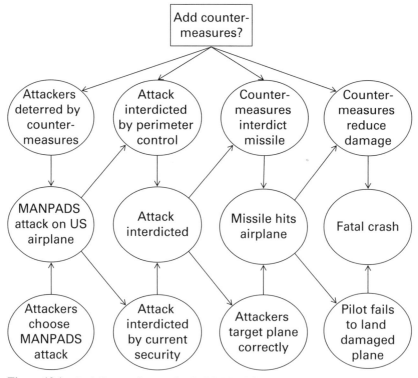

Figure 10.1   An influence diagram for the MANPADS countermeasure decision.

about whether the countermeasures will be effective if they are put into place. The events represented by the top row of uncertainties have zero probability if the decision is to not implement countermeasures. von Winterfeldt and O'Sullivan parameterized the probabilities for the top row and bottom row of uncertainties and allowed decision-makers to determine their own probabilities and perform sensitivity analysis. This was necessary because the events represent complex combination of more specific events and are thus difficult for intelligence experts to assess. For instance, the probability of an attempt involves a number of potential terrorist groups, their ability to obtain MANPADS, and their ability to get the MANPADS into the United States and near a target. Note that the attacker's decision to attempt a MANPADS attack is treated as an uncertainty by von Winterfeldt and O'Sullivan.

Bakir (2008) performed a decision analysis for decisions about screening containers entering the United States by truck over the Mexican border for radioactive material or devices that could be used in an attack on targets in the United States. Three decisions are considered: improving transportation security, installing screening equipment at the Mexican border, and whether to use Advanced Scintillator Portals (ASPs) or keep existing Radiation Portal Monitors (RPMs). Merrick and McLay (2010) extend Bakir's analysis, examining specifically if screening should be performed at all. Figure 10.2 shows an influence diagram for this

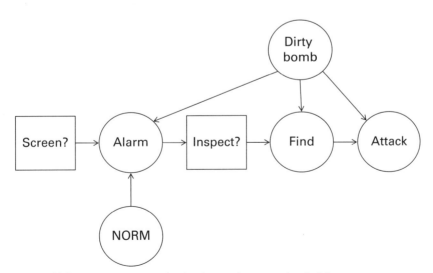

Figure 10.2   An influence diagram for the container screening decision.

decision that includes whether to inspect the container if the device indicates radioactive material in the container. Naturally occurring radioactive material (NORM) sources are frequently shipped in containers, so such alarms can often not be a threat. The probability of an alarm signal must take into account whether terrorists have obtained material for a dirty bomb and decide to smuggle it into the United States in a container, as well as the probability of NORM sources inside the container. Observe again that the attacker decision to put a dirty bomb in the container is treated as an uncertainty by Merrick and McLay.

Decision analysis has also been used to rank critical infrastructures (Haimes, Kaplan, and Lambert, 2002; Apostolakis and Lemon, 2005), to determine bridges to protect (Leung, Lambert, and Mosenthal, 2004), and to treat victims of a radioactive attack (Feng and Keller, 2006). The representation of uncertainty in such counterterrorism decisions frequently involves probability elicitation from intelligence experts, each of whom is required to survey large amounts of intelligence data. One of the most difficult uncertainties to model is the probability of different terrorist actions and how the terrorists will adapt to defender actions. Merrick and McLay (2010) examine the limits of probabilistic modeling of adaptation through changes in threat probabilities. They point out that the judgment task is complex, leading to probability assessments that may be wrong. (In technical parlance, such assessments may be "miscalibrated," meaning that, e.g., events that are judged to happen with at least a certain frequency may not, in reality, do so.) Brown and Cox (2010) question the value of expert judgments and the use of nonadaptive models of terrorist behavior, given the attacker's ability to game the resulting recommendations. Merrick and Parnell (2010) review adaptive and nonadaptive approaches to counterterrorism decisions and conclude that each can be useful when the required assumptions are met—but that what assumptions reflect reality is often in doubt.

A different stream of development applies data mining to intelligence information, avoiding the need to speculate about or model adversarial decision processes and instead emphasizing detecting adversarial plans and behaviors. However, the data is not structured conventionally, in terms of predefined fields and values, but arises as text, speech, photos and videos, Internet browsing patterns, cell phone and social media usage patterns, and so forth. The large volume of such "unstructured" data is beyond the capabilities of most human processors. However, advances in natural language processing allow the categorization and prioritization of unstructured data for human processing or even automated extraction of information.

The remainder of this chapter reviews these three alternative approaches. First, we discuss eliciting probabilities of terrorist actions. Next, we review the use of adaptive decision and game theory to model terrorist actions and reactions. Lastly, we review the use of natural language processing and text mining to analyze intelligence information and determine likelihoods of terrorist actions. Arguably, these different approaches represent the past, present, and future of advanced decision and risk analysis. Since the 1960s, part of the canon of decision analysis has been subjective expected utility (SEU) theory, including elicitation and calibration of subjective probabilities. The limitations of human expertise have gradually been clarified by psychologists (as discussed in Nobel Prize winner Daniel Kahneman's recent book *Thinking Fast and Slow*) and by social scientists who have studied the performance of expert probability judgments (e.g., Tetlock, 1995). Expert elicitation methods face many of the same key practical challenges today (especially that experts probability judgments are often wrong) than they did half a century ago. By contrast, modeling of intelligent adversaries and agents has progressed dramatically, with a variety of insights from behavioral decision theory and game theory illuminating what real people actually *do*, as opposed to what idealized rational agents *should* do. Modeling the real behavior of intelligent, emotional people brings a new layer of realism and predictive power to game-theoretic modeling. Finally, the detection of predictively useful patterns and integration of more knowledge than human experts can handle, distilled from a wide variety of unstructured data, represents the future (and parts of the present) of "big data" and unstructured data processing technologies as they apply to improving risk analysis capabilities. The integration of these approaches holds great promise for improved risk analysis and better-informed decisions for managing risks of complex systems, especially when the actions of other human beings drive key risks and uncertainties. Cyber threats, financial system stability, national and international supply chains, political unrest, organized crime, and other challenging areas of social, political, and economic risk may soon benefit from these techniques.

ELICITING ATTACKER ACTIONS FROM EXPERTS

Ezell et al. (2010) assert that the probabilities of different attacker alternatives should be elicited from intelligence experts. The use of expert judgment is not without its difficulties. In a practical sense, the main concern in using expert judgment in place of data is the validity of the judgments provided. Are the probabilities provided actually a valid scale related to the frequency of the events (Wallsten and Budescu, 1983)? Calibration is one measure of validity. When the expert assesses the probability P does the event occur $P\%$ of the time? The decision-maker would hope that P should be close to the observed relative frequency of the event, $\hat{p}$. However, P is really a measure of the expert's degree of belief and may not be based on the same data as $\hat{p}$ (Keren, 1997). Most literature observes the well-known S-shaped calibration curve in practice, caused by probability judgments that are more extreme than the corresponding relative frequency.

Calibration is not the only attribute of a good probability judgment. Consider the case of an expert assessing the probability that a container will set off the RPM alarm. An expert can say 2.5% and be perfectly calibrated, as this is the observed relative frequency of NORM sources across all containers, but this does not provide any helpful information about terrorist threats in a specific container (Liberman and Tversky, 1993). Alternatively, an expert can always say there is a threat when there is not and say there is no threat when there is. The expert is then perfectly miscalibrated but highly informative as you can just assume the opposite of what the expert says (since P is related to $\hat{p}$ in a known way).

What mechanisms affect the calibration and sharpness of probability judgments? When faced with the task of considering complex events such as these, we tend to simplify the task using simpler heuristic mechanisms. These heuristics have been shown to generate systematic biases in our judgments. We search our memory for relevant knowledge and combine that knowledge to form a judgment (Wallsten and Budescu, 1983). The judgment then depends on the quality of our search for information and our ability to combine information to make a judgment. Errors can be introduced as little effort is often used for these cognitive processes (Hora, 2007). Three main heuristics are discussed in the literature, although there are many more: the representativeness heuristic, the availability heuristic, and the anchoring heuristic. We will discuss each in turn in the context of the aforementioned examples.

When asked to assess the probability that an individual is a terrorist or that a container contains a threatening device, we may use the representativeness heuristic. We evaluate the probability by "the degree to which A is representative of B, that is, the degree to which A resembles B" (Tversky and Kahneman, 1974). Does the individual resemble a terrorist, or does this container seem like a threat? We compare what we are told about the subject to a typical member of the category of interest. The probability assessments are based on the representativeness of the individual or container, and the overall base rates of terrorists or threats are not considered. This can lead to racial profiling or overestimation of threats in containers.

We often judge the frequency of events by our ease of recalling similar events. This is called the availability heuristic (Kahneman and Tversky, 1973). For example, we may underestimate the probability of a MANPADS attack if we have not previously seen such an event. However, external events like the media can also lead to overestimation (Combs and Slovic, 1979). It also depends on how closely one experiences a traumatic event. Kahneman, Slovic, and Tversky (1982) show that seeing an accident will lead to better recall than learning about it secondhand through a newspaper. The same would appear to be true for a terrorist incident. Fischoff et al. (2003) find a tendency to overestimate the probability of a terrorist event in people that live close to a previous event, with the tendency being more pronounced in males, whites, adults, and Republicans.

To simplify the judgment task, we often look for a starting value and then adjust away from that value. However, the starting value, which can come from the formulation of the question or from initial computations by the expert, can become an anchor. Experts often insufficiently adjust away from initial anchors (Clemen and Reilly, 2001). This heuristic is called anchoring and adjustment (Tversky and Kahneman, 1974). The bias is revealed when we change the anchor and get different judgments in repeated elicitations. For example, when we ask an expert for a best estimate and then a range, they will often adjust insufficiently away from the best estimate, making the range too small. This makes our judgments overconfident, as we do not admit to sufficient uncertainty.

Ravinder, Kleinmuntz, and Dyer (1988) and Howard (1989) show that decomposing complex events (breaking the problem down into smaller or more manageable chunks) improves the calibration and sharpness of probability judgments. This helps when we are thinking about how the event in question relates to or is caused by other events (Clemen and Reilly, 2001). Decomposition should ideally provide a breakdown that is easier to contemplate. Ravinder, Kleinmuntz, and Dyer (1988) assessed the reliability of judgments achieved with varying levels of decomposition. They found that increasing the level of decomposition does reduce random errors in the assessment, but only up to a point, and that each component event should be chosen to be as close as possible to the observable history of the expert to improve the accuracy of the probability judgments. Andradóttir and Bier (1997) find that it is better to err on the side of too many component events rather than too few to avoid significant drop-offs in performance.

Returning to our examples, we should note that von Winterfeldt and O'Sullivan (2006), Bakir (2008), and Merrick and McLay (2010) all acknowledged the potential for errors in probability assessments. Thus, they used ranges of probabilities and performed extensive sensitivity analysis to determine whether decisions would be affected by mis-specification. This is always a hallmark of good decision analysis. However, even with such sensitivity analysis, assessing the probabilities of terrorist actions is not without its difficulties and errors.

USING ADAPTIVE DECISION AND GAME THEORY

Game theory applications in counterterrorism explicitly model the decision of the terrorist (Kunreuther and Heal, 2003; Bier, 2005; Heal and Kunreuther, 2005; Banks and Anderson, 2006; Zhuang and Bier, 2007; Zhuang et al., 2007; Bier et al., 2007a,b). Paté-Cornell and Guikema (2002), Parnell, Smith, and Moxley (2010), and Rios Insua, Jesus, and Banks (2009) combined game-theoretic and decision analysis approaches for counterterrorism. The aim of this stream of research is to work prescriptively to allocate defender resources in a way that minimizes risks from terrorist attacks.

Let us consider a single example from the perspective of these different approaches. Figure 10.3 shows a representative decision from the perspective of the defender. As an example of this setup, suppose the defender must decide whether to screen for nuclear threats in containers. Given this decision, the attacker may or may not obtain the material and capability for a radioactive dispersal device (RDD) and

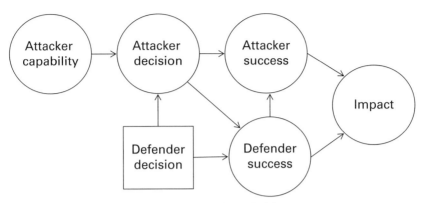

Figure 10.3   An influence diagram for the defender decision.

then must decide whether to use this capability or to launch a more conventional attack. If the attacker decides on the RDD, they must get it through the defender's screening (if any) and then successfully carry out the attack. If they decide on a conventional attack, then the screening will obviously not be effective. If the attack is successful, then there are varying levels of impact depending on which type of attack the attacker chose. As we are using a defender influence diagram, an uncertainty node represents the attacker decision (as observed previously in von Winterfeldt and O'Sullivan (2006) and Merrick and McLay (2010)). This probability must be elicited for each such node, as discussed in Section "Eliciting attacker actions from experts". To solve the defender decision, we perform backward induction in Figure 10.3 by taking expected consequences at each uncertainty node and minimizing expected consequences at the defender's decision node.

Instead, we may choose to apply the red team approach and model the decision from the attacker's perspective, as in Figure 10.4. The attacker decision is now a decision node. However, we now need two such diagrams, one when the defender decides to screen and one for no screening. To solve each attacker decision, we perform backward induction by taking expected consequences at each uncertainty node and maximizing expected consequences at the attacker decision node.

To avoid this complexity, we may choose to use sequential game theory and explicitly model both the defender and attacker decision nodes, as in Figure 10.5. To save space, we show only the uncertainty node for the marginal distribution of the impact given the attacker and defender decisions. To solve for both decisions, we perform backward

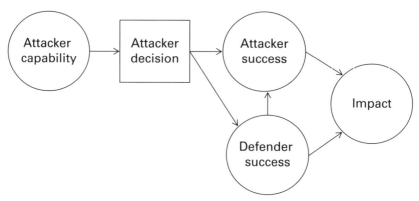

Figure 10.4 An influence diagram for the attacker decision.

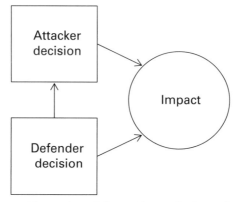

Figure 10.5 A sequential game theory influence diagram for the attacker–defender decision.

induction by taking expected consequences for each possible combination of attacker and defender choices and then maximizing expected consequences at the attacker decision node and minimizing expected consequences at the defender decision node. However, as shown in Figure 10.6, not all uncertainty nodes are after the two decision nodes, so Figure 10.5 does not represent the correct chronological order.

Recent methods proposed by Parnell, Smith, and Moxley (2010) and Rios Insua, Jesus, and Banks (2009) solve for the attacker decision. The solution method finds the optimal decision for the attacker and the defender, where both are faced with uncertainty about the outcomes. Their approaches are similar to game theory, but the focus is prescriptive (helping the defender find the optimal decision), rather than trying

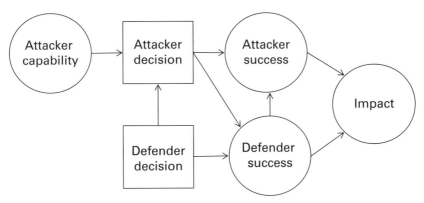

Figure 10.6   An influence diagram using intelligent adversary risk analysis.

to descriptively model the interplay between attackers and defenders (as in an economic or political study of counterterrorism). Figure 10.6 shows the approach proposed by Parnell, Smith, and Moxley (2010), called "intelligent adversary risk analysis." Unlike Figure 10.5, the defender decision, attacker decision, and uncertainties are all represented in their correct time sequence. To solve the intelligent adversary risk analysis influence diagram, we perform backward induction by taking expected consequences for uncertainty nodes, maximizing expected consequences at the attacker decision node, and minimizing expected consequences at the defender decision node. Parnell et al. combine game theory and decision analysis to model sequential problems. In game theory parlance, Parnell et al.'s approach is a zero-sum game with common and perfect information.

The decision-maker may also choose to model the probabilities and objectives of the attacker and defender as different and may have uncertainty about the specific values the attacker may use. Rios Insua, Jesus, and Banks (2009) propose joint attacker and defender decision diagrams, an approach they call "adversarial risk analysis." The attacker decision is solved by backward induction as in Figure 10.4. The anticipated result of the attacker decision, occurring later in time than the defender decision, is used as an input to the defender influence diagram. In Figure 10.7, we show the defender's uncertainty about the attacker's beliefs, leading to a different solution to the attacker decision for each possibility. This induces a distribution over the attacker's decision that is used as an attacker uncertainty node in the defender decision. The defender decision is then solved by backward induction as in Figure 10.3.

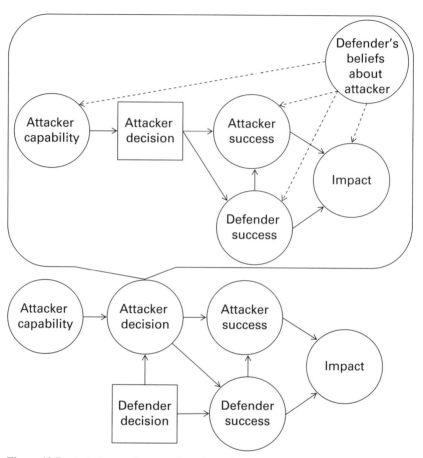

Figure 10.7 An influence diagram using adversarial risk analysis with defender uncertainty about attacker beliefs.

As noted earlier, Ezell et al. (2010) specify a probability distribution over the set of alternatives faced by the attacker as in Figure 10.3. The adversarial risk analysis ends up with such a distribution as an output of the attacker model, but only elicits inputs to this model not outputs like Ezell et al., so the judgments will be better calibrated as discussed in Section "Eliciting attacker actions from experts."

Each of the models discussed thus far makes assumptions about rationality of the decision-making process. In particular, the defender decision model assumes that the defender is rational and follows the foundations of SEU (Ramsey, 1931; Savage, 1954). Classical game theory models assume all decision-makers are rational (von Neuman and Morgenstern, 1947). Behavioral game theory models do not assume rationality and attempt to represent what people actually do when faced with such strategic interactions with other decision-makers. Parnell, Smith, and Moxley (2010) assume that all decision-makers are rational and share the same probabilities and utilities. One could think of this as modeling what the defender would do if faced with the attacker's decision. Rios Insua, Jesus, and Banks (2009) assume both that the defender is rational and that the defender believes the attacker is rational and can express her beliefs about the attacker's probabilities and utilities. In fact, both Parnell, Smith, and Moxley (2010) and Rios Insua, Jesus, and Banks et al. (2009) use expected consequences in their examples and do not employ utility functions. Zhuang and Bier (2007) do consider the attacker's risk aversion by explicitly including a utility function in a game-theoretic setup. Wang and Bier (2011) extend the modeling of the attacker's decision to consider a multiattribute utility function for selecting a target. Wang and Bier (2013) extend this work to statistically determine the attributes of a target from expert judgments from intelligence officials, thus combining the approaches of Sections "Eliciting attacker actions from experts" and "Using adaptive decision and game theory."

It is well understood that without the use of decision analytic methods, human beings are not purely rational (while also not behaving in completely irrational or random ways) (Kahneman and Tversky, 1974). Heuristics are employed to aid in decision-making, as exhibited repeatedly in experiments on decision behavior. These lead to deviations from the behavior predicted by SEU. Descriptive decision models have been developed to represent actual decision-maker behavior for risky decisions with a single decision-maker. For example, real decision-makers will change their choices regarding gambles depending on how they are described. These are known as framing effects and go against the axioms of rationality (Tversky and Kahneman, 1981). We are averse to losses, rather than considering just our end point, and are overly attached to what we already have (endowment effect), rather than being willing to swap current and future endowments based on their utilities (Kahneman, 1991). We are also overly attracted to the sure thing (certainty effect). Descriptive decision models like prospect theory (Kahneman and Tversky, 1979; Tversky and Kahneman, 1992) can replicate these effects and are thus good predictors of observed decision behavior.

In counterterrorism, if we are attempting to model attacker decisions, we should not necessarily model them as purely rational. Shan and Zhuang (2013) introduce a probability that an attacker will act in a nonstrategic (nonexpected utility) manner. If the attacker does act strategically, then their action is determined using expected utilities and standard game theory approaches. If the attacker does not act strategically, then their action is assigned randomly from a predetermined distribution. This mixture distribution of rational and random actions is a good first step to reduce the pure rationality assumption. However, solving for the optimal attacker decision using SEU implies that we are ignoring biases that humans exhibit in making decisions under uncertainty without prescriptive help, like gain–loss effects, endowment effects, certainty effects, and probability neglect. The deviation from expected utility is better understood than simply assigning random actions. This could mean that our models of attacker behavior are biased or just plain wrong. As we are trying to help the defender make better decisions, errors in our predictions of attacker behavior can in turn lead to defender decisions that are not truly optimal. Instead, we should ideally represent attacker decisions with descriptive decision models (Kahneman and Tversky, 1979; Tversky and Kahneman, 1992; Birnbaum, 2005).

Such work is only beginning in the game theory literature but shows promise to improve adversarial decision and game theory. Behavioral economics and behavioral game theory study such deviations in multiple decision-maker situations. As pointed out by Camerer (2003), "behavioral game theory is about what players actually do." For instance, a fully rational player should iteratively delete all dominated strategies. However, experimental evidence shows that people stop after a few such iterations (Nagel, 1995; Hoffman, McCabe, and Smith, 1996; Ho, Camerer, and Weigelt, 1998). The longer people play a given game, the more levels of iteration they consider and the closer they get to an equilibrium solution. However, they will not necessarily avoid the deviations from expected utility found in prospect theory and other descriptive models of decision under risk and uncertainty.

Shalev (2000) borrowed prospect theory's central concept of reference dependence, where decision-makers value outcomes not in an overall sense, but as gains and losses from a reference point. The most notable example of this effect is loss aversion: individuals assign greater weight to outcomes perceived as losses relative to their reference point than to outcomes perceived as gains. Shalev aimed to incorporate this concept of loss aversion in a natural way into the traditional game-theoretic solution concept of Nash equilibrium. Metgzer and Rieger (2010) addressed the effects of probability neglect by incorporating nonlinear probability weights into traditional game-theoretic solution concepts. Metzger and Rieger (2010) defined a new solution concept: mixed-strategy prospect-theoretic equilibrium. However, Shalev (2000) and Metzger and Rieger (2010) use prospect theory only to model uncertainty about the other players' actions. We should also seek to incorporate these ideas in the combined decision analysis and game theory approaches of Parnell, Smith, and Moxley (2010) and Rios Insua, Jesus, and Banks (2009) to provide more accurate representations of the attacker's behavior in these counterterrorism risk and decision models.

NATURAL LANGUAGE PROCESSING TO DETERMINE TERRORIST INTENT

Each of the methods discussed thus far aims to understand the likelihood of potential terrorist actions. The first approach requires intelligence analysts to consume and filter the vast amount of potentially relevant information and then form judgments about the probabilities and uncertainties in the decision model. The second approach seeks to model the terrorist's decision directly but still requires judgments about the terrorist's beliefs and preferences. Paté-Cornell (2002) discusses the need for "fusion" of intelligence information (the process of merging "the content of signals, some sharp and some fuzzy, some independent and others not, into useful information"). Paté-Cornell proposes the use of Bayesian methods for this fusion, where the prior on a hypothesized terrorist action or plan is updated using the assessed likelihood of the observed signals were that hypothesis be true, instead of some other potential hypothesis. Figure 10.8 shows a Bayesian network by Paté-Cornell, where messages are observed indicating different stages of the terrorist attack chain. The main problem is the scale of this process, as there are large numbers of signals, and many of these signals are in the form of unstructured data such as text or speech.

Natural language processing is the use of computers to obtain formal representations of meaning from human or natural language sources (Manning and Schutze, 1999; Liddy, 2003). The first

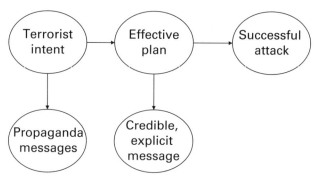

Figure 10.8 A Bayesian network of intelligence fusion from Paté-Cornell (2002).

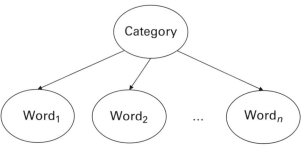

Figure 10.9 A Bayesian network of a bag of words.

applications of such methods aimed to classify or categorize documents so they could be assigned appropriately to various analysts (Yilmazel et al., 2002). This process is similar to automated website tagging or even information retrieval in library science. Numerous methods have been developed, including the naive Bayes classifier (McCallum and Nigam, 1998), latent semantic indexing (Deerwester et al., 1990; Hofmann, 1999), and support vector machines (Joachims, 2002). The underlying concept behind many of these techniques is the "bag of words" (Dumais et al., 1998; Sebastiani, 2002), where the frequency of a word is assumed to be independent of the frequencies of other words, given a category, classification, or tag (Fig. 10.9). These methods can be used to categorize research articles as biology, chemistry, or physics, but they can also be used to classify emails, texts, and phone calls as terrorist related or not. The likelihood of a given set of word frequencies is then estimated for each possible category from a corpus or training set of documents. The frequency of words is calculated for each document in the corpus, and the relevant categories are defined by experts in the subject matter.

The algorithm is trained using the corpus and can then be used to give the probability of each potential category for a particular document. These algorithms can be used either supervised or unsupervised. In a supervised application, the predicted categories can be verified by a subject matter expert and then the document added to the corpus.

The next level of natural language processing is information extraction (Bikel, Schwartz, and Weischedel, 1999; Cowie and Wills, 2000; Doddington et al., 2004; Boschee, Weischedel, and Zamanian, 2005), taking text in any form and generating structured information (Zamin, 2009). Figure 10.10 shows the general model form for information extraction, where words are distributed based on underlying concepts and the concepts follow an underlying process. To attempt to discover meaning in text, one must first preprocess the data to cleanse it of images and unwanted white spaces, to tokenize the text, and to tag parts of speech. This process is common to all natural language processing, and again, multiple methods have been proposed, including hidden Markov models (Merialdo, 1994), rule-based algorithms (Sleator and Temperley, 1993), probabilistic context-free grammars (Klein and Manning, 2003), and maximum entropy methods (Toutanova and Manning, 2000). Once the parts of speech have been tagged, there are three general steps to achieve information extraction: named entity recognition, coreference resolution, and entity extraction. Named entity recognition classifies entities such as the names of people, organizations, locations, dates, and so on (Riloff, 1993; Kim and Moldovan, 1995; Madhyastha, Balakrishnan, and Ramakrishnan, 2003). Coreference resolution resolves which entities are referred to across different parts of text: when are references to the

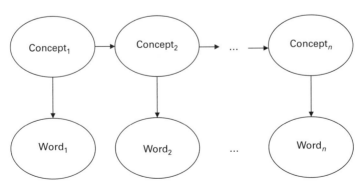

Figure 10.10   A Bayesian network of a hidden Markov model for information extraction.

same entity and when are they different? Common techniques include maximum entropy methods (Versley et al., 1998) and hidden Markov models (Soon, Lim, and Ng, 1999). Finally, entity extraction algorithms identify relevant facts in text, classifying them into predefined categories. This process can use support vector machines (Sun et al., 2003) or pattern learning algorithms (Kwee et al., 2005).

Once information has been extracted from text data, the next step is to make connections between entities across different documents. Social network analysis looks at social relationships in terms of network or graph theory. These techniques have been used to analyze text on the web (Kuramochi and Karypis, 2002; Schenker et al., 2004, 2005; Cox and Holder, 2007) but can be more specifically applied to understanding a terrorist network (Diesner and Carley, 2005; Airoldi et al., 2007). Figure 10.11 shows a social graph of the 9–11 terrorists, developed manually, at a scale much smaller than is now possible with automated processes. This work starts with information extraction and then uses link analysis, the detection of potentially interesting links between entities that are supported by the corpus of data (Rosa, 2004; Badia and Kantardzic, 2005). Krebs (2002) and Zamin (2009) each independently represent the connections between the 9 and 11 attackers as a link analysis as represented in Figure 10.11.

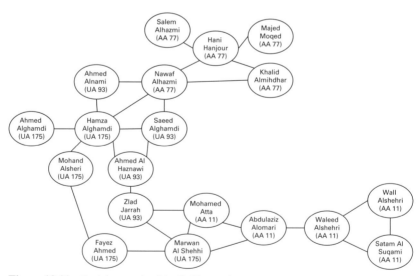

Figure 10.11   Social network of the 9–11 terrorists.

The final step in using natural language processing in analysis of terrorism documents and websites is intent extraction. The goal of intent extraction is to determine the presence of threats or planned attacks (Singh et al., 2006). Pattipati et al. (2006) use hidden Markov models and Bayesian networks to assess the presence of a planned attack, while Weinstein et al. (2009) use support vector machines applied to the output of the information extraction and link analysis. The difficulty in developing working prototypes is the need for a corpus of data consisting of terrorist communications and text from situations that did lead up to an attack and also from situations that did not lead up to an attack. Blaylock and Allen (2005) develop a simulated corpus based on historical terrorist attacks, with controllable true signals and noise. With a corpus to train these models, they mirror Paté-Cornell's approach in Figure 10.8, but the likelihoods are estimated from the corpus rather than by intelligence experts.

Natural language processing has great potential in counterterrorism. It can be used as a means to prioritize data sources (Yilmazel et al., 2002) or to directly extract information and links to inform intelligence analysts (Rosa, 2004; Badia and Kantardzic, 2005). The analysts are then better prepared for the probability elicitations required by the approaches discussed in Sections "Eliciting attacker actions from experts" and "Using adaptive decision and game theory." Alternatively, it can be used directly to estimate probabilities for use in counterterrorism decision models (Singh et al., 2006; Pattipati et al., 2006; Weinstein et al., 2009). In either mode, automated processing of unstructured data will improve the detection of relevant signals. Given the imperfect nature of such methods, this process is best performed in a supervised mode, keeping the intelligence analyst in the process.

CONCLUSIONS

We have reviewed three approaches to determining the likelihood of terrorist actions: direct probability elicitation, adaptive decision and game theory, and natural language processing. While each of these approaches has its benefits and challenges, a combination of the three approaches is likely to be the best way forward. We have seen that adaptive decision and game theory allows a meaningful decomposition of complex events into more directly quantifiable elements. This improves the calibration of probability judgments but leaves the intelligence analyst with a vast amount of unstructured data to review. The inclusion of natural language processing in this process can improve the timeliness of judgments, the ability to include all relevant signals in the data, and the sharpness of the probability judgments. Thus, all three approaches combined have the promise to yield better-calibrated and sharper judgments and hence better-informed risk management decisions. If we then model attacker decisions using descriptive decision theory, we can remove an additional source of bias and achieve a more solid foundation for counterterrorism decisions.

REFERENCES

Airoldi, E.M. D.M., Blei, S.E., Fienberg, A. Goldenberg, E. Xing, A. Zheng. eds. 2007. *Statistical Network Analysis: Models, Issues, and New Directions: ICML 2006 Workshop on Statistical Network Analysis*, Pittsburgh, PA, USA, June 29, 2006, Lecture Notes in Computer Science, Vol. 4503, Springer-Verlag.

Andradóttir, S., V. M. Bier. 1997. Choosing the number of conditioning events in judgemental forecasting. *Journal of Forecasting* **16** 255–286.

Apostolakis, G. E., D. M. Lemon. 2005. A screening methodology for the identification and ranking of infrastructure vulnerabilities due to terrorism. *Risk Analysis* **25**(2) 361–376.

Badia, A., M. Kantardzic. 2005. Link analysis tools for intelligence and counter terrorism. *Lecture Notes in Computer Science*, **3495**, 49–59.

Bakir, N. O. 2008. A decision tree model for evaluating countermeasures to secure cargo at United States southwestern ports of entry. *Decision Analysis* **5**(4) 230–248.

Banks, D., S. Anderson. 2006. Game theory and risk analysis in the context of the smallpox threat. In *Statistical Methods in Counterterrorism*, A. Wilson, G. Wilson, and D. Olwell, eds., Springer, New York, 9–22.

Bier, V. M. 2005. Game-theoretic and reliability methods in counter-terrorism and security. In *Mathematical and Statistical Methods in Reliability, Series on Quality, Reliability and Engineering Statistics*, A. Wilson, N. Limnios, S. Keller-McNulty, Y. Armijo, eds., World Scientific, Singapore, 17–28.

Bier, V. M., E. R. Gratz, N. J. Haphuriwat, W. Magua, K. R. Wierzbicki. 2007a. Methodology for identifying near-optimal interdiction strategies for a power transmission system. *Reliability Engineering & System Safety* **92** 1155–1161.

Bier, V. M., S. Oliveros, L. Samuelson. 2007b. Choosing what to protect: Strategic defensive allocation against an unknown attacker. *Journal of Public Economic Theory* **9**(4) 563–587.

Bikel, D. M., R. L. Schwartz, R. M. Weischedel. 1999. An algorithm that learns what's in a name. *Machine Learning* **34** 1–3.

Birnbaum, M. H. 2005. Three new tests of independence that differentiate models of risky decision making. *Management Science* **51**(9) 1346–1358.

Blaylock, N., J. Allen. 2005. Generating artificial corpora for plan recognition. In *Lecture Notes in Artificial Intelligence*, 3938, L. Ardissono, P. Brna, A. Mitrovic, eds. Springer, Edinburgh, 179–188.

Boschee, E., R. Weischedel, A. Zamanian. 2005. *Automatic Information Extraction*. In Proceedings of the 2005 International Conference on Intelligence Analysis, McLean, VA, May 2–4, 2005.

Branting, L.K. 2005. *Name Matching in Law Enforcement and Counter-Terrorism*, ICAIL Workshop on Data Mining, Information Extraction, and Evidentiary Reasoning for Law Enforcement and Counter-Terrorism, Bologna, Italy June 11, 2005.

Brown, G. G., L. A. Cox. 2010. How probabilistic risk assessment can mislead terrorism risk analysis. *Risk Analysis* **31**(2) 196–204.

Camerer, C. 2003. *Behavioral Game Theory: Experiments in Strategic Interaction*, Princeton University Press, Princeton, NJ.

Clemen, R., T. Reilly. 2001. *Making Hard Decisions*, Duxbury Press, Pacific Grove, CA.

Combs, B., P. Slovic. 1979. Newspaper coverage of causes of death. *Journalism Quarterly* **56**(4) 837–843.

Cook, D. J., L. B. Holder. 2007. *Mining Graph Data*, John Wiley & Sons, Inc., Hoboken, NJ.

Cowie, J., Y. Wills. 2000. *Information Extraction, A Handbook of Natural Language Processing*, Marcel Dekker, New York.

Deerwester, S., S. Dumais, T. Landauer, G. Furnas, R. Harshman. 1990. Indexing by latent semantic analysis. *Journal of the American Society of Information Science* **41**(6) 391–407.

Diesner, J., K. Carley. 2005. *Exploration of Communication Networks from the Enron Email Corpus*. In Proceedings of the Workshop on Link Analysis, Counterterrorism and Security, SIAM International Conference on Data Mining 2005, Society for Industrial and Applied Mathematics, Philadelphia, April 21–23, 3–14.

Doddington, G., A. Mitchell, M. Pryzbocki, L. Ramshaw, S. Strassel, R. Weischedel. 2004. *The Automatic Content Extraction (ACE) Program Tasks, Data, and Evaluation*. In Proceeding of LREC 2004 Conference on Language resources and Evaluation, Lisbon, May 24–30.

Dumais, S., J. Platt, D. Heckerman, M. Sahami. 1998. *Inductive Learning Algorithms and Representations for Text Categorization*. In Proceedings of the Seventh International Conference on Information and Knowledge Management, Washington, DC, USA, ACM Press, Bethesda, MD, November 2–7, 1998.

Ezell, B.C., S.P. Bennett, D. von Winterfeldt, J. Sokolowski, A.J. Collins. 2010. Probabilistic risk analysis and terrorism. *Risk Analysis* **30**(4) 575–589.

Feng, T., L. R. Keller. 2006. A multiple-objective decision analysis for terrorism protection: Potassium iodide distribution in nuclear incidents. *Decision Analysis* **3**(2) 76–93.

Fischoff, B., R. M. Gonzalez, D. A. Small, J. S. Lerner. 2003. Judged terror risk and proximity to the world trade center. *Journal of Risk and Uncertainty* **26**(2/3) 137–151.

Haimes, Y. Y., S. Kaplan, J. H. Lambert. 2002. Risk filtering, ranking, and management framework using hierarchical holographic modeling. *Risk Analysis* **22**(2) 383–397.

Heal, G., H. Kunreuther. 2005. IDS models of airline security. *Journal of Conflict Resolution* **49**(2) 201–217.

Ho, T., C. Camerer, K. Weigelt. 1998. Iterated dominance and iterated best-response in experimental "p-beauty contests." *American Economic Review* **88** 947–969.

Hofmann, T. 1999. *Probabilistic Latent Semantic Indexing.* In SIGIR '99 Proceedings of the 22nd Annual International ACM SIGIR Conference on Research and Development in Information Retrieval, pages 50–57, University of California, Berkeley, August 15–19, 1999, ACM, New York, NY.

Hoffman, E., K. McCabe, V. L. Smith. 1996. On expectations and monetary stakes in ultimatum games. *International Journal of Game Theory* **25** 289–301.

Hora, S. 2007. Eliciting probabilities from experts. In *Advances in Decision Analysis: From Foundations to Applications,* W. Edwards, R.F. Miles, D. von Winterfeldt, eds., Cambridge University Press, Cambridge, 129–153.

Howard, R.A. 1989. Knowledge maps. *Management Science* **35**(8) 903–922.

Joachims, T. 2002. *Learning to Classify Text using Support Vector Machines,* Kluwer Academic Publishers, Boston, MA.

Kahneman, D., A. Tversky. 1973. On the psychology of prediction. *Psychology Review* **80**(4) 237–251.

Kahneman, D., A. Tversky. 1974. Judgment under uncertainty: Heuristics and biases. *Science* **185**(4157) 1124–1131.

Kahneman, D. 1991. The endowment effect, loss aversion, and status quo bias: Anomalies. *Journal of Economic Perspectives* **5**(1) 193–206.

Kahneman, D., A. Tversky. 1979. Prospect theory: An analysis of decision under risk. *Econometrica* **47** 263–292.

Kahneman, D., P. Solvic, A. Tversky. 1982. *Judgment under Uncertainty: Heuristics and Biases,* Cambridge University Press, Cambridge.

Keren, G. 1997. On the calibration of probability judgments: Some critical comments on alternate perspectives. *Journal of Behavioral Decision Making* **10** 269–278.

Kim, J., D. Moldovan. 1995. Acquisition of Linguistic Patterns for Knowledge-based Information Extraction. *IEEE Transactions on Knowledge and Data Engineering* **7** 713–724.

Klein, D., C.D. Manning. 2003. Fast exact inference with a factored model for natural language parsing. In *Advances in Neural Information Processing Systems* 15 (NIPS 2002), S. Becker, S. Thrun, K. Obermayer, eds., MIT Press, Cambridge, MA, 3–10.

Krebs, V. E. 2002. Uncloaking terrorist networks. *First Monday* **7**(4).

Kunreuther, H., G. Heal. 2003. Interdependent security. *Journal of Risk & Uncertainty* **26**(2–3) 231–249.

Kuramochi, M., G. Karypis. 2002. An Efficient Algorithm for Discovering Frequent Subgraphs, Technical Report TR# 02-26, Department of Computer Science and Engineering, University of Minnesota.

Kwee, O.T. L.E. Peng, R. Gunaratna, S. Zhen. 2005. *Event Driven Document Selection for Terrorism Information Extraction.* In Proceedings of the 3rd IEEE International Conference of Intelligence and Security Informatics, Springer-Verlag, Berlin, Heidelberg.

Leung, M., J. H. Lambert, A. Mosenthal. 2004. A risk-based approach to setting priorities in protecting bridges against terrorist attacks. *Risk Analysis* **24**(4) 963–984.

Liberman, V., A. Tversky. 1993. On the evaluation of probability judgments. *Psychological Bulletin* **114**(1) 162–173.

Liddy, E. D. 2003. *Natural Language Processing. Encyclopedia of Library and Information Science*, Marcel Decker, Inc., New York.

Madhyastha, H. V., N. Balakrishnan, K. R. Ramakrishnan. 2003. *Event Information Extraction Using Link Grammar.* In Proceedings of the 13th International Workshop on Multi-lingual Information Management, Hyderabad, India, March 10–11, 2003, IEEE, Piscataway, NJ, 16–22.

Manning, C., H. Schutze. 1999. *Foundations of Statistical Natural Language Processing*, MIT Press, Cambridge, MA.

McCallum, A., K. Nigam. 1998. *A Comparison of Event Models for Naive Bayes Text Classification*, AAAI–98 Workshop on Learning for Text Categorization, Melbourne, Australia, August 24–28, 1998.

Merialdo, B. 1994. *Tagging english text with a probabilistic* model. *Computational Linguistics* **20**(2) 155–171.

Merrick, J. R. W., L. A. McLay. 2010. Is screening cargo containers for smuggled nuclear threats worthwhile? *Decision Analysis* **7**(2) 155–171.

Merrick, J. R. W., G. S., Parnell. 2010. A comparative analysis of PRA and intelligent adversary methods for counterterrorism risk management. *Risk Analysis* **31**(9) 1488–1510.

Metgzer, L. P., M. O. Rieger. 2010. Equilibria in games with prospect theory preferences. Unpublished paper.

Nagel, R. 1995. Unraveling in guessing games: An experimental study. *American Economic Review* **85** 1313–1326.

Parnell, G. S., C. M. Smith, F. I. Moxley. 2010. Intelligent adversary risk analysis: A bioterrorism risk management model. *Risk Analysis* **30**(1) 32–48.

Paté-Cornell, E. 2002. Fusion of intelligence information. *Risk Analysis* **22**(3) 445–454.

Paté-Cornell, E., S. Guikema. 2002. Probabilistic modeling of terrorist threats: A systems analysis approach to setting priorities among countermeasures. *Military Operations Research* **7**(4) 5–20.

Pattipati, K. R., P.K. Willett, J. Allanach, H. Tu, S. Singh. 2006. Hidden markov models and Bayesian networks for counter-terrorism. In *Emergent Information Technologies and Enabling Policies for Counter Terrorism*, R. Popp, J. Yen, eds., John Wiley & Sons, Inc., Hoboken, NJ, 27–50.

Ramsey, F. P. 1931. Truth and probability. In *The Foundations of Mathematics and Other Logical Essays*, F. P. Ramsey, ed., Harcourt, Brace, New York, 156–198.

Ravinder, H.V., D.N. Kleinmuntz, J. Dyer. 1988. The reliability of subjective probabilities obtained through decomposition. *Management Science* **34**(2) 186–199.

Riloff, E. 1993. *Automatically Constructing a Dictionary for Information Extraction Tasks.* In Proceedings of the 11th National Conference on Artificial Intelligence, Washington Convention Center, Washington, D.C., July 11–15, 1993, Association for the Advancement of Artificial Intelligence, 811–816.

Rios Insua, D., J. Jesus, D. Banks. 2009. Adversarial risk analysis. *Journal of the American Statistical Association* **104**(486) 841–854.

Rosa, M. D. 2004. *Data Mining and Data Analysis for Counter Terrorism*, Center for Strategic and International Studies Report, CSIS Press, Washington, DC.

Savage, L. J. 1954. *Foundations of Statistics*, John Wiley & Sons, Inc, New York.

Schenker, M. L., H. Bunke, A. Kandel. 2004. Classification of web documents using graph matching. *International Journal of Pattern Recognition and Artificial Intelligence* **18**(3) 475–496.

Schenker, A., H. Bunke, M. Last, A. Kandel. 2005. Graph-theoretic techniques for web content mining. *World Scientific, Series in Machine Perception and Artificial Intelligence* **62**.

Sebastiani, F. 2002. Machine learning in automated text categorization. *ACM Computing Surveys* **34**(1) 1–47.

Shalev, J. 2000. Loss aversion equilibrium. *International Journal of Game Theory* **29**(2) 269–287.

Shan, X., J. Zhuang. 2013. Hybrid defensive resource allocations in the face of partially strategic attackers in a sequential defender–attacker game. *European Journal of Operational Research* **228** 262–272.

Singh, S., W. Donat, J. Lu, K. Pattipati, P. Willett. 2006. *An Advanced System for Modeling Asymmetric Threats*. In IEEE International Conference on Systems, Man, and Cybernetics, Taipei, Taiwan, October.

Sleator, D.D., D. Temperley. 1993. *Parsing English with A Link Grammars*. In 3rd International Workshop on Parsing Technologies (ACL – SIGPARSE), University of Tilburg, the Netherlands August 10–13, 1993.

Soon, W.M., D.C.Y. Lim, H.T. Ng. 2001. A machine learning approach to coreference resolution of noun phrases. *Computational Linguistics* **27**(4) 521–544.

Sun, A., M. M. Naing, E. P. Lim, W. Lam. 2003. Using support vector machine for terrorism information extraction. *Lecture Notes in Computer Science* **2665** 1–12.

Tetlock, P. E. 1995. *Expert Political Judgment: How good is it? How can we know?* Princeton University Press, Princeton, NJ.

Toutanova, K., C. D. Manning. 2000. Enriching the knowledge sources used in a maximum entropy part-of-speech tagger. Proceedings of the Joint SIGDAT Conference on Empirical Methods in Natural Language Processing and Very Large Corpora (EMNLP/VLC-2000), pp. 63–70, Hong Kong, October 7–8, 2000, Hong Kong University of Science and Technology (HKUST), Hong Kong, The Association for Computational Linguistics, New Brunswick.

Tversky, A., D. Kahneman.1974. Judgment under uncertainty: Heuristics and biases. *Science* **185** 1124–1131.

Tversky, A., D. Kahneman. 1981. The framing of decisions and the psychology of choice. *Science* **21**(4481) 453–458.

Tversky, A., D. Kahneman. 1992. Advances in prospect theory: Cumulative representation of uncertainty. *Journal of Risk and Uncertainty* **5**(4) 297–323.

Versley, Y., S. P. Ponzetto, M. Poesio, V. Eidelman, A. Jern, J. Smith, X. Yang, A. Moschitti. 2008. BART: A Modular Toolkit for Coreference Resolution. In Companion Volume of the Proceedings of the 46th Annual Meeting of the Association for Compuatational Linguistics (ACL 2008), pp. 9–12. June 15–20, 2008, The Ohio State University, Columbus, Ohio, USA, The Association for Computational Linguistics, New Brunswick, NJ.

von Neuman, J., O. Morgenstern. 1947. *Theory of Games and Economic Behavior*, Princeton University Press, Princeton, NJ.

von Winterfeldt, D., T. M. Sullivan. 2006. Should we protect commercial airplanes against surface-to-air missile attacks by terrorists? *Decision Analysis* **3**(2) 63–75.

Wallsten, T.S., D.V.Budescu. 1983. Encoding subjective probabilities: A psychological and Psychometric review. *Management Science* **29**(2) 151–173.

Wang, C., V. M. Bier. 2011. Target-hardening decisions based on uncertain multiattribute terrorist utility. *Decision Analysis* **8**(4) 86–302.

Wang, C., V. M. Bier. 2013. Expert elicitation of adversary preferences using ordinal judgments. *Operations Research* **61**(2) 372–385.

Weinstein, C., W. Campbell, B. Delaney, G. O'Leary. 2009. Modeling and Detection Techniques for Counter-Terror Social Network Analysis and Intent Recognition. Proceedings of the IEEE Aerospace Conference, pp. 1–16, March 07–14, 2009, Big Sky Resort, Big Sky, MT, IEEE, Piscataway, NJ.

Yilmazel, O., S. Symonenko, N. Balasubramanian, E. D. Liddy. 2002. *Improved Document Representation for Classification Tasks for the Intelligence Community.* In Proceedings of the AAAI Stanford Spring Symposium American Association for Artificial Intelligence, March 25–27, 2002 in Palo Alto, CA, Association for the Advancement of Artificial Intelligence, Palo Alto, CA.

Zamin, N. 2009. Information Extraction for Counter-Terrorism: A Survey. *Computation World* '09, November 15–20, 2009, pp. 520–526, Athens, IEEE, Piscataway, NJ.

Zhuang, J., V. M. Bier. 2007. Balancing terrorism and natural disasters—Defensive strategy with endogenous attacker effort. *Operations Research* **55**(5) 976–991.

Zhuang, J., V.M. Bier, A. Gupta. 2007. Subsidies in interdependent security with heterogeneous discount rates. *The Engineering Economist* **52**(1) 1–19.

Index

Breakthroughs in Decision Science and Risk Analysis, First Edition.
Edited by Louis Anthony Cox, Jr.
© 2015 John Wiley & Sons, Inc. Published 2015 by John Wiley & Sons, Inc.